CONVERSION FACTORS

1 in.	= 2.54 cm (Exactly)
1 mi.	= 1.609 km
1 lb.	= 0.45359237 kg (Exactly)
1 AMU	= 1.6605×10^{-27} kg
(1 AMU) $\times c^2$	= 931.481 MeV
30 mi/hr	= 44 ft/sec
1 eV	= 1.602×10^{-19} J
1 MeV	= 1.602×10^{-13} J
1 Cal	= 4186 J
1 tesla (T)	= 10^4 gauss (G)

ASTRONOMICAL DATA

1 Light year (L.Y.)	= 9.46×10^{15} m
1 Astronomical unit (A.U.) (Earth-Sun distance)	= 1.50×10^{11} m
Radius of Sun	= 6.96×10^8 m
Earth-moon distance	= 3.84×10^8 m
Radius of Earth	= 6.38×10^6 m
Radius of moon	= 1.74×10^6 m
Mass of Sun	= 1.99×10^{30} kg
Mass of Earth	= 5.98×10^{24} kg
Mass of moon	= 7.35×10^{22} kg
Average orbital speed of Earth	= 2.98×10^4 m/sec ($\cong$ 30 km/sec)

PHYSICAL CONSTANTS

Velocity of light in vacuum	$c = 2.998 \times 10^8$ m/sec
Charge of electron	$e = 1.60 \times 10^{-19}$ C
Planck's constant	$h = 6.63 \times 10^{-33}$ J-sec
Avogadro's number	$N_o = 6.02 \times 10^{23}$ mole^{-1}
Electron mass	$m_e = 9.11 \times 10^{-31}$ kg
	$m_e c^2 = 0.511$ MeV
Proton mass	$m_p = 1.6726 \times 10^{-27}$ kg
	$= 1.007276$ AMU
	$= 1836.11 \ m_e$
	$m_p c^2 = 938.26$ MeV
Neutron mass	$m_n = 1.6749 \times 10^{-27}$ kg
	$= 1.008665$ AMU
	$m_n c^2 = 939.55$ MeV
Gravitational constant	$G = 6.673 \times 10^{-11}$ N-m^2/kg^2

PHYSICS
THE FOUNDATION OF MODERN SCIENCE

PHYSICS
THE FOUNDATION OF
MODERN SCIENCE

Jerry B. Marion

Department of Physics and Astronomy
University of Maryland

John Wiley & Sons, Inc. New York · London · Sydney · Toronto

This book was set in Times Roman with display in Futura by York Graphic Services, Inc., and printed and bound by Halliday Lithographers, Inc. The designer was Nancy Gruber. The drawings were designed and executed by the Wiley Illustration Department. Joan E. Rosenberg supervised production. Cover art by Jerome Wilke. Photo, Magnum.

Library of Congress Cataloging in Publication Data:

Marion, Jerry B.

  Physics: the foundations of modern science.

  1. Physics.   I. Title.

QC21.2.M365   530   72-10175
ISBN 0-471-56917-8

Printed in the United States of America

10 9 8 7 6 5 4 3 2 1

PREFACE

Today's world is a technological world. One cannot escape this fact; the evidence is all around us, in almost every facet of modern life. And tomorrow's world—the world in which today's students will spend most of their lives—will be based even more broadly on a foundation of science and technology. How are we to cope with the increasing reliance on scientific concepts and complex technical equipment in our everyday lives? It is hopeless to understand the world today—much less to be able to influence and direct the world of tomorrow—without some appreciation of science and the scientific basis of modern technology.

It is to this end that this book has been designed. The science of *physics,* which deals with the fundamental aspects of the behavior of matter and energy has provided the foundation stones on which rest most of the other scientific and technological disciplines. The fields of chemistry, meteorology, engineering, geology, oceanography, even modern biology and medicine draw heavily on physical principles as they progress. It is the purpose of this book to provide an overview of physics with emphasis on the modern ideas that have influenced developments in other scientific fields.

This book is directed toward the large segment of college and university students who have no particular training in science. What should a student expect to derive from a course of study based on this book? The mathematical level of the discussions here is undemanding—only some familiarity with high-school algebra is required. And although solving some elementary numerical exercises is helpful in gaining an appreciation for the way physics *works,* problem solving is not the main objective of this book. Instead, the presentation is *concept oriented* and is designed to show the broad range of physical ideas and how they are interrelated. Physics is not a collection of isolated facts. It is a highly interconnected organic body, growing and changing as new facts are discovered and as new ideas are generated. It is the goal of this book to dispel the conception of physics as an "impossible subject" and to provide an insight into the structure, the beauty, and the importance of science.

Jerry B. Marion

CONTENTS

PHYSICS
THE FOUNDATION OF MODERN SCIENCE

CHAPTER 1
PHYSICS AND THE UNIVERSE AROUND US

In our Universe, Man finds himself in a middle position. He is enormously larger than the atoms and molecules that make up his body and his surroundings, and yet his size is insignificantly small in comparison with the stars and galaxies that comprise the Universe. Even more remarkable than the existence of this vast range of distances and sizes in the Universe is the fact that Man, from his middle position, has found ways to study both the microscopic domain of atoms and molecules and the cosmic realm of galaxies and space. These studies have led to the discovery of many of the natural laws that govern the behavior of our Universe and its constituent parts. What we have accomplished and what we are still attempting to learn—this is the theme of this book.

There is a strong coupling of physics to the world in which we live. The pursuit of knowledge about the behavior of Nature leads to discoveries that ultimately affect our everyday lives. Most scientific investigations are concerned only with deepening our understanding of Nature. But the results of these investigations frequently find applications for the benefit (or detriment) of mankind. Indeed, our modern society—for better or for worse—rests on a foundation of technology that is fostered by science. It is the aim of this book to provide a brief look at the underpinning of the house in which we live.

1.1 *Physics as an Experimental Science*

THE FINAL TEST IS IN THE LABORATORY

The scientist seeks to learn the "truth" about Nature. In physics we can never learn "absolute truth" because physics is basically an experimental science; experiments are never perfect and, therefore, our knowledge of Nature must always be imperfect. We can only state at a certain moment in time the extent and the precision of our knowledge of Nature, with the full realization that both the extent and the precision will increase in the future. Our understanding of the physical world has as its foundation experimental measurements and observations; on these are based our theories that organize our facts and deepen our understanding.

Physics is not an armchair activity. The ancient Greek philosophers debated the nature of the physical world, but they would not test their conclusions, they would not experiment. Real progress was made only centuries later, when man finally realized that the key to scientific knowledge lay in observation and experiment, combined with logic and reason. Of course, the generation of ideas in physics involves a certain amount of just plain *thinking,* but when the final analysis is made, the crucial questions can only be answered in the laboratory.

THE PHILOSOPHY OF DISCOVERY

The mere accumulation of facts does not constitute good science. Certainly, facts are a necessary ingredient in any science, but facts alone are of limited value. In order to fully utilize our facts, we must understand the relationships among them; we must systematize our information and discover how one event produces or influences another event. In doing this, we follow the *scientific method:* the coupling of observation, reason, and experiment.

The scientific method is not a formal procedure or a detailed map for the exploration of the unknown. In science we must always be alert to a new idea and prepared to take advantage of an unexpected opportunity. Progress in science occurs only as the result of the symbiotic relationship that exists between observational information and the formulation of ideas that correlate the facts and allow us to appreciate the interrelationships among the facts. The scientific method is actually not a "method" at all; instead, it is an attitude or philosophy concerning the way in which we approach the real physical world and attempt to gain an understanding of the way Nature works.

Johannes Kepler (1571–1630), the greatest of the early astronomers, followed the scientific method when he analyzed an incredible number of observations of the positions of planets in the sky. From these facts he was able to deduce the correct description of planetary motion: the planets move in elliptical orbits around the Sun.

Kepler's procedure—amassing facts and trying various hypotheses until he found one that accounted for all the information—is not the only way to utilize the scientific method. When Erwin Schrödinger was working on the problems associated with the new experiments in atomic physics in the 1920s, he set out to find a description of atomic events that could be formulated in a mathematically beautiful way. Instead of closely following the experimental facts and attempting to relate them, he sought only to find an aesthetically pleasing mathematical description of the general trend of the results. This pursuit of mathematical beauty led Schrödinger to develop modern quantum theory. In the realm of atoms, where quantum theory applies, Nature does indeed operate in a beautiful way.

We cannot impose any rigid constraints on the development of science. Different individuals work in different ways. As long as we combine logic and experiment we follow the scientific method.

PHYSICS AND
THE UNIVERSE
AROUND US

2

THE LANGUAGE OF PHYSICS

One of the significant steps forward in Man's understanding of the behavior of Nature was the realization that it is the Earth that moves around the Sun, and not the Sun that moves around the Earth. The simple statement that "the Earth moves around the Sun" represents a new dimension in physical thinking. As important as this idea is, nevertheless, it is incomplete. We cannot say that we really understand a physical phenomenon until we have reduced the description to a statement involving *numbers*. Physics is a precise science and its natural language is mathematics. Only when Johannes Kepler gave a mathematical description of planetary motion and Isaac Newton derived the same results on the basis of his theory of universal gravitation could it be said that a proper analysis of the motion of the Earth and the planets had finally been made.

In this book the emphasis is on the conceptual basis of physics. But we will still require *some* mathematics to pursue the discussion in a meaningful way. It will be necessary to quote numbers from time to time and to express certain ideas in equation form. However, we will not burden the reader with more mathematical details than seem absolutely necessary.

THE "POWERS-OF-10" NOTATION

The physical world encompasses things very small and things very large—the size of an atom and the size of the Universe, the time required for light to travel from this page to your eye and the time that the Universe has existed. In dealing with these enormous ranges of distance and time (and other quantities), we encounter an annoying problem of notation. For example, if we wish to express the ratio of the diameter of a dime to the distance from the Earth to the Sun, we can write this as

$$\frac{\text{diameter of dime}}{\text{Earth-Sun distance}} = \frac{1}{15,000,000,000,000}$$

or, in decimal notation,

$$\frac{\text{diameter of dime}}{\text{Earth-Sun distance}} = 0.000\ 000\ 000\ 000\ 067$$

Obviously, neither of these methods of expressing the result is particularly convenient. To overcome this difficulty, we use the "powers-of-10" notation in which the number of times that 10 is multiplied together appears in the result as the superscript of 10 (called the *exponent* of 10 or the *power* to which 10 is raised). That is, $10 \times 10 = 100 = 10^2$ and $10 \times 10 \times 10 = 10^3$. Clearly, $10^1 = 10$, and, by convention, $10^0 = 1$.

Products of powers of 10 are expressed as

$$10^2 \times 10^3 = (10 \times 10) \times (10 \times 10 \times 10) = 10^5 = 10^{(2+3)}$$

That is, in general, the product of 10^n and 10^m is $10^{(n+m)}$:

$$10^n \times 10^m = 10^{(n+m)} \tag{1.1}$$

If the power of 10 appears in the denominator, the exponent is given a negative sign:

$$\frac{1}{10} = 0.1 = 10^{-1} \qquad \text{and} \qquad \frac{1}{1000} = 0.001 = \frac{1}{10^3} = 10^{-3}$$

In general,

$$\frac{1}{10^m} = 10^{-m} \tag{1.2}$$

Usually, we express numbers in this notation by writing the coefficient of the power of 10 as a number between 0.1 and 10. Thus, we write either 2.4×10^9 or 0.24×10^{10} instead of, for example, 24000×10^5 or 0.00024×10^{13}.

Using the powers-of-10 notation, we can now express the ratio of the diameter of a dime to the Earth-Sun distance as 6.7×10^{-14}. Expressed in this form, the information is much more easily assimilated than in the previously used form involving 13 zeroes.

Example **1.1**

$$\frac{6,400,000}{400} = (6.4 \times 10^6) \times \frac{1}{4 \times 10^2}$$

$$= \frac{6.4}{4} \times \frac{10^6}{10^2} = 1.6 \times 10^4$$

When discussing physical quantities, we sometimes use a prefix to the unit instead of the appropriate power of 10. For example, *centi-* means $\frac{1}{100}$ so that $\frac{1}{100}$ of a meter is called a *centimeter; mega-* means 10^6 so that $\$1,000,000 = 1$ megabuck. Table 1.1 lists some of the most frequently used prefixes.

EXPLANATION OF SYMBOLS

In physics we are often concerned with the way in which two (or more) physical quantities are related. The *distance* that an automobile travels, for example, depends on its *speed* and on the *time*. The most convenient and economical way to express the relationship between physical quantities is frequently in the form of an *equation*. If we wish to know how far a dropped ball will fall in a certain period of time, we find that we can express the relationship between the length of fall y (in feet) and the time t (in seconds) as

$$y = 16t^2$$

If we know a result only approximately, but not exactly, we use the symbol $\cong$, with the meaning *is approximately equal to*. Thus, if the *exact* expression relating y and t^2 were

$$y = 16.127t^2$$

Table **1.1** *Prefixes Equivalent to Powers of 10*

Prefix	Symbol	Power of 10
giga-	G	$10^{9\,a}$
mega-	M	$10^{6\,a}$
kilo-	k	10^{3}
centi-	c	10^{-2}
milli-	m	10^{-3}
micro-	μ	10^{-6}
nano-	n	10^{-9}
pico-	p	10^{-12}
femto-	f	10^{-15}

$^{a}\,10^{6} = 1$ *million.* In the US, $10^{9} = 1$ *billion,* but the European convention is that $10^{9} = 1000$ million and that 1 billion $= 10^{12}$; the prefix *giga-* is internationally agreed on to represent 10^{9}.

we could write approximate relations as

$$y \cong 16.13\, t^{2} \qquad \text{or} \qquad y \cong 16.1\, t^{2} \qquad \text{or} \qquad y \cong 16\, t^{2}$$

The use of approximations of this type (for example, replacing 16.127 by 16) is a respectable procedure and may even be preferred to the "exact" calculation in many instances. It depends on how precise we wish our final result to be. There is no sense in carrying out a calculation to six decimal places if two will do. It is well to remember the maxim, "that which is good enough is best."

If we have only a vague idea of the magnitude of a quantity, then we use the symbol $\approx$ or $\sim$ to imply that the number is only *very* approximately known or *is of the order of magnitude of* some number. For example, we say that the galaxy Andromeda is about 3×10^{11} as massive as the Sun. But we do not know this number very precisely; it could be 2×10^{11}, or 6×10^{11}, or even farther from 3×10^{11}. Therefore, we say: *The mass of Andromeda is of the order of magnitude of* 3×10^{11} *times the mass of the Sun,* and we write

mass of Andromeda $\approx 3 \times 10^{11} \times$ mass of the Sun

or we say that the ratio of the mass of Andromeda to the mass of the Sun is $\sim 3 \times 10^{11}$.

The symbols $>$ and $<$ mean, respectively, *is greater than* and *is less than:*

area of New York $<$ area of Texas
mass of the Earth $>$ mass of the moon

If a certain quantity is *very much larger than* another quantity, we use the symbol $\gg$:

mass of Andromeda $\gg$ mass of Sun

The symbols that we will find useful in this book are listed in Table 1.2 along with their meanings.

Table **1.2** *Mathematical Symbols and Their Meanings*

Symbol	Meaning
$=$	is equal to
$\propto$	is proportional to
$\cong$	is approximately equal to
$\approx$ or $\sim$	is *very* approximately equal to; is of the order of magnitude of
$>$ ($<$)	is greater (less) than
$\gg$ ($\ll$)	is much greater (less) than

1.3 *The Fundamental Units of Measure*

UNITS AND STANDARDS

In our subsequent discussions we will encounter a variety of physical quantities—for example, length, time, mass, force, momentum, energy, and so forth. These quantities not only have magnitudes but they have *units* or *dimensions* as well. It makes no sense to state that a certain length is 12—we must also specify the units in which the magnitude has this value. Whether the length is 12 centimeters or 12 miles makes a considerable difference!

It is also necessary to have *standards* for the units of physical measure. If we state that the size of a building lot is 30 paces by 60 paces, we have only a crude idea of the area. But if we state that the size is 20 meters by 40 meters, we know the area precisely because the *meter* is a well-defined and standard unit of length. In this book we will use the *metric system* of physical measure. But because the United States has not yet converted to this international system and clings (temporarily) to the British system, we will make a gradual transition to the metric system by using both systems in the first few chapters.

THE STANDARD OF LENGTH

Although man has invented an enormous number of units for the specification of length, only those of the British and metric systems survive today. The unit of length in the British system is the *yard;* the derived units are the *foot* ($\frac{1}{3}$ yard), the *inch* ($\frac{1}{36}$ yard), and the statuate mile (1760 yards). In the metric system the unit of length is the *meter,* which was originally conceived as 10^{-7} of the distance from the equator to the North Pole along a meridian passing through Paris. In order to provide a more practical standard, in 1889 the meter was officially defined as the distance between two parallel scribe marks on a certain bar of platinum-iridium.

The meter-bar definition of the meter suffers from two disadvantages: not only is the precision inadequate for many scientific purposes, but comparisons of lengths with a bar which is kept in a standards laboratory are quite inconvenient. These difficulties have been overcome with the definition, by international agreement in 1961, of a *natural* unit of length based on an atomic radiation. Because all atoms of a given species are identical, their

Fig. 1.1 *A krypton-86 lamp is adjusted in its liquid nitrogen bath. The orange-red light from such lamps is used as the international standard of length.*

radiations are likewise identical. Therefore, an atomic definition of length is reproducible everywhere. We now accept as the standard of length the wavelength (in vacuum) of a particular orange radiation emitted by krypton gas (see Fig. 1.1). The standard was arrived at by carefully measuring the length of the standard meter bar in terms of the wavelength of krypton light. It was then decided that exactly 1,650,763.73 wavelengths would constitute 1 meter. This definition is then consistent with the previous definition in terms of the distance between the scribe marks on the meter bar, but it has the advantage of being approximately 100 times as precise. Now the standard can be reproduced in many laboratories throughout the world, not in standards laboratories alone.

Table 1.3 gives the conversion factors connecting some of the units of

Table **1.3** *Conversion Factors for Length*

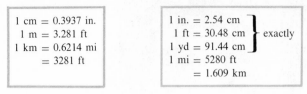

1 cm = 0.3937 in.	1 in. = 2.54 cm
1 m = 3.281 ft	1 ft = 30.48 cm ⎫
1 km = 0.6214 mi	1 yd = 91.44 cm ⎬ exactly
= 3281 ft	1 mi = 5280 ft ⎭
	= 1.609 km

Table **1.4** *The Range of Distances in the Universe*

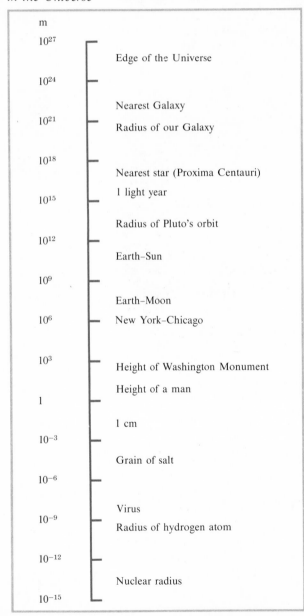

m

10^{27} — Edge of the Universe

10^{24} —

— Nearest Galaxy

10^{21} — Radius of our Galaxy

10^{18} —

— Nearest star (Proxima Centauri)

10^{15} — 1 light year

— Radius of Pluto's orbit

10^{12} —

— Earth–Sun

10^{9} —

— Earth–Moon

10^{6} — New York–Chicago

10^{3} —

— Height of Washington Monument

— Height of a man

1 —

— 1 cm

10^{-3} —

— Grain of salt

10^{-6} —

— Virus

10^{-9} — Radius of hydrogen atom

10^{-12} —

— Nuclear radius

10^{-15} —

length in the metric and British systems. Notice that the inch, the foot, and the yard are now defined *exactly* in terms of centimeters. Thus, the krypton wavelength is the standard of length for both systems.

Table 1.4 shows in a schematic way the enormous range of distances we encounter in the Universe. Notice that there is a factor of 1000 between successive marks on the vertical scale. Between the smallest and the largest things about which we have any comprehension, the span is more than 40 factors of 10!

Table 1.5 gives the values of some of the distances that we will find useful in our discussions.

Table **1.5** *Some Useful Distances*

To Proxima Centauri (nearest star)	4.04×10^{16} m
1 light year (L.Y.)	9.460×10^{15} m
1 astronomical unit (A.U.) (Earth–Sun distance)	1.496×10^{11} m
Radius of Sun	6.960×10^{8} m
Earth–Moon distance	3.844×10^{8} m
Radius of Earth	6.378×10^{6} m
Wavelength of yellow sodium light	5.89×10^{-7} m
1 Ångstrom (Å)	10^{-10} m
Radius of hydrogen atom	5.292×10^{-11} m
Radius of proton	1.2×10^{-15} m

THE STANDARD OF TIME

The unit of time in both the British and metric systems is the *second*, which until recently was defined as $\frac{1}{86,400}$ of the mean solar day. The determination of time by observation of the rotation of the Earth is inadequate for high precision work because of minor but quite perceptible anomalies in the speed of the Earth's rotation.

Table **1.6** *Relative Precision of Various Types of Clocks*

	Precision	
Type of Clock	*1 sec in*	*1 part in*
Hour glass	1.5 min	10^{2}
Pendulum clock	3 hr	10^{4}
Tuning fork	1 day	10^{5}
Quartz crystal oscillator	3 yr	10^{8}
Ammonia resonator	30 yr	10^{9}
Cesium resonator	3×10^{4} yr	10^{12}
Hydrogen maser	3×10^{6} yr	10^{14}

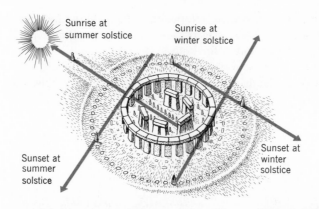

Sunrise at summer solstice

Sunrise at winter solstice

Sunset at summer solstice

Sunset at winter solstice

Fritz Henle—Monkmeyer

Fig. 1.2 *An ancient time-telling device—the Stonehenge calendar, in southern England, about 3500 years old. By standing in the center of the monument at the altar stone, the high priest could see the rising Sun through the pillars on the first day of summer. The arrows drawn on the drawing above show the alignment of the various landmarks that pointed to the rising and setting Sun on the days of the summer and winter solstices. Only the sunrise at the summer solstice is viewed from inside the monument thus suggesting that this was a particularly significant day, perhaps the first day of the new year.*

In order to improve the precision of time measurements, we adopted in 1967 a *natural* unit for time, just as we had done previously for a length standard. Our present day *atomic clocks* depend on the characteristic vibrations of cesium atoms. The second is defined as the time required for 9,192,631,770 cycles of the particular atomic vibrations in cesium. With this definition of the second, it is possible to compare time intervals to 1 part

PHYSICS AND
THE UNIVERSE
AROUND US

in 10^{12}, which corresponds to 1 second in 30,000 years. Current research with other atomic vibrations (notably those of the hydrogen maser) indicates that we will soon have a clock that will be accurate to 1 part in 10^{14}, or to 1 second in 3 million years!

Table 1.7 shows the range of time intervals that we encounter in the Universe. Notice that the span from the shortest to the longest time interval is greater than 40 orders of magnitude, about the same as the range of distances shown in Table 1.4.

Table **1.7** *The Range of Time Intervals in the Universe*

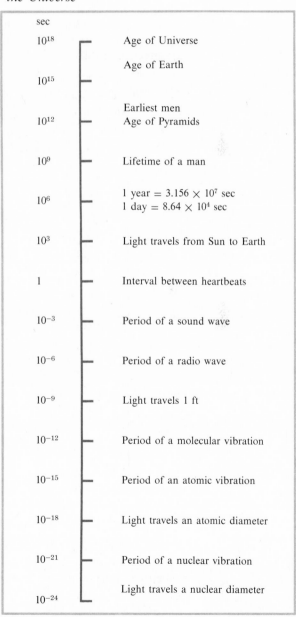

sec	
10^{18}	Age of Universe
	Age of Earth
10^{15}	
10^{12}	Earliest men
	Age of Pyramids
10^{9}	Lifetime of a man
10^{6}	1 year = 3.156×10^7 sec
	1 day = 8.64×10^4 sec
10^{3}	Light travels from Sun to Earth
1	Interval between heartbeats
10^{-3}	Period of a sound wave
10^{-6}	Period of a radio wave
10^{-9}	Light travels 1 ft
10^{-12}	Period of a molecular vibration
10^{-15}	Period of an atomic vibration
10^{-18}	Light travels an atomic diameter
10^{-21}	Period of a nuclear vibration
	Light travels a nuclear diameter
10^{-24}	

THE STANDARD OF MASS

The international standard of mass is a cylinder of platinum-iridium[1] which is defined as 1 kilogram = 10^3 grams. In the British system the unit of mass is the *pound:* 1 lb = 0.45359237 kg.

It would, of course, be highly desirable to have an atomic standard for mass just as we have for length and time. We do have such a standard which is used in the comparison of the masses of atoms and molecules, but, unfortunately, we have no precision method at present of utilizing this standard above the level of individual atoms and molecules. When the technology has developed to the point that we can determine precisely the mass of the standard kilogram in terms of the atomic mass standard, we will certainly adopt the atomic unit of mass as our standard.

Table 1.8 shows some of the marker points on the gigantic range of masses that we find in the Universe. Values of some of the more important masses are given in Table 1.9.

Table **1.8** *The Range of Masses in the Universe*

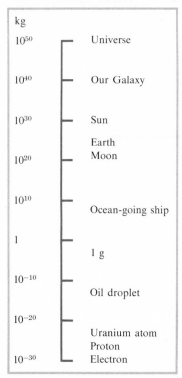

kg	
10^{50}	Universe
10^{40}	Our Galaxy
10^{30}	Sun
	Earth
10^{20}	Moon
10^{10}	
	Ocean-going ship
1	
	1 g
10^{-10}	
	Oil droplet
10^{-20}	
	Uranium atom
	Proton
10^{-30}	Electron

CONSERVATION OF MASS

Mass is the first *property of matter* (as distinct from the basically geometrical concepts of length and time) that we have introduced. We have come to appreciate the permanence and the immutability of mass. A block

[1]Originally, the kilogram was defined as the mass of 1000 cm^3 of water.

Table **1.9** *Some Important Masses*

Object	Mass (kg)
Sun	1.991×10^{30}
Earth	5.977×10^{24}
Moon	7.35×10^{22}
Proton	1.672×10^{-27}
Electron	9.108×10^{-31}

of metal has a certain mass. This mass does not change with time. Nor does it change when its dimensions are changed; we can alter its shape by hammering or forging, but the mass does not change. We can even dissolve it in acid, but the combined mass of the acid and the metal (together with any gases that might be evolved in the process) remains unchanged.

These facts concerning mass constitute the first of a series of important fundamental statements called the *conservation laws* of physics. Instead of emphasizing the *differences* between various physical processes, in order to understand the fundamentals of Nature it is more profitable to seek those properties that are the *same* in any process. The conversion of water into ice, and the conversion of water into its constituent gases, oxygen and hydrogen, are two different physical processes. But a common feature of these processes is that the *mass* of the material remains constant. A long series of experiments, all of which exhibit this feature, has led us to conclude that *mass is conserved* in all physical processes. This is the first of several *conservation laws* that we will discuss.[2]

DENSITY

How do we determine the mass of an object that is too large to place on a scale? It is a simple matter if we know the volume of the object. First, we take a small sample of the material of which the object is composed and measure its volume and mass. The ratio of these quantities determines the *density* of the material:

$$\text{density} = \frac{\text{mass}}{\text{volume}}$$

$$\rho = \frac{M}{V} \tag{1.3}$$

Once the density has been determined, the mass of the large object is given by the product of the density and the volume. If the object is not homogeneous (that is, if the density is not the same everywhere in the object), then an *average* density must be found before the mass can be determined.

In the metric system, densities are expressed either in kg/m^3 or in g/cm^3 (see Table 1.10).

---

[2] As we will see in later chapters, relativity theory prescribes a relationship connecting mass and energy. Therefore, the conservation law properly refers, not to mass alone, but to *mass-energy*. However, in all everyday processes the distinction is not important, and we can treat mass as a conserved quantity.

Table **1.10** *Some Representative Densities*

Type of Mattter	Density g/cm³	Density kg/m³
Nuclear matter	10^{14}	10^{17}
Center of Sun	10^2	10^5
Lead	11.3	1.13×10^4
Aluminum	2.7	2.7×10^3
Water	1	10^3
Air	10^{-3}	1
Laboratory high vacuum	10^{-18}	10^{-15}
Interstellar space	10^{-24}	10^{-21}
Intergalactic space	10^{-30}	10^{-27}

Example **1.2**

What is the average density of the Earth?

The radius of the Earth is 6.38×10^6 m. Therefore, the volume is

$V = \frac{4}{3}\pi R^3 = \frac{4}{3}\pi \times (6.38 \times 10^6\,m)^3$
$= 1.09 \times 10^{21}\,m^3$

The mass of the Earth is 5.98×10^{24} kg, so the average density is

$$\rho = \frac{M}{V} = \frac{5.98 \times 10^{24}\,\text{kg}}{1.09 \times 10^{21}\,\text{m}^3}$$
$$= 5.5 \times 10^3\,\text{kg/m}^3 \text{ (or 5.5 g/cm}^3)$$

This is the density averaged over the entire Earth; actually, the core has a density of $\sim 12 \times 10^3$ kg/m³ and the mantle (the region near the surface) has a density of $\sim 3 \times 10^3$ kg/m³.

THE USE OF UNITS

All physical quantities have *units* or *dimensions*. When we make numerical statements or write numerical equations concerning physical quantities, we must include the units of the quantities. If we make the statement, "The distance traveled is equal to the speed multiplied by the time," we could express this more briefly as

distance = speed × time

PHYSICS AND
THE UNIVERSE
AROUND US

or, by giving arbitrary symbols to the quantities, as

$d = s \times t$

This equation does not explicitly contain the units of the various quantities; the equation is valid for any system of units as long as the system is used consistently. For example,

$$30 \text{ mi} = 15 \frac{\text{mi}}{\text{hr}} \times 2 \text{ hr}$$

Not only must the *numbers* balance in such an equation, but so must the *units*. On the right-hand side of the equation, the time unit "hour" occurs in both numerator and denominator and therefore cancels, leaving "miles" as the unit on both sides of the equation.

We can alter the unit of any quantity by using *conversion factors*. In order to convert 15 mi/hr to the corresponding number of feet/second, we use the relationships

$$1 \text{ hr} = 3600 \text{ sec} \qquad \text{and} \qquad 1 \text{ mi} = 5280 \text{ ft}$$

Then, we can form the ratios

$$\frac{1 \text{ hr}}{3600 \text{ sec}} = 1 \qquad \text{and} \qquad \frac{5280 \text{ ft}}{1 \text{ mi}} = 1$$

Because we can multiply any quantity by a factor of *unity* without altering the value (we only change the *scale*), we can write

$$15 \frac{\text{mi}}{\text{hr}} = 15 \frac{\cancel{\text{mi}}}{\cancel{\text{hr}}} \times \frac{1 \cancel{\text{hr}}}{3600 \text{ sec}} \times \frac{5280 \text{ ft}}{1 \cancel{\text{mi}}}$$

$$= \frac{15 \times 5280}{3600} \frac{\text{ft}}{\text{sec}}$$

$$= 22 \frac{\text{ft}}{\text{sec}}$$

Note how "miles" and "hours" have cancelled in this expression.

1.4 *Man's Place in the Universe*

THE ASTRONOMICAL STANDARD OF LENGTH

Only a few hundred years ago the Earth was believed to be the center and the most important feature of the Universe. Today we understand how astronomically insignificant our Earth really is: we live on a medium-size planet that orbits around a medium-size star that lies in an outer region of a medium-size galaxy. How have we come to appreciate the relationship of the Earth to the Universe? What measurements have we made to establish the distances to stars and the extent of the Universe?

Essentially all astronomical determinations of distance are based on a standard of length that is the distance from the Earth to the Sun. This length standard is called the *astronomical unit* (A.U.). When Kepler analyzed the motions of the planets (particularly Mars) in the 17th century, he was able to express planetary distances only in terms of the Earth-Sun distance. With modern telescopes and recording techniques we can determine planetary distances in A.U. with extremely high precision. But how do we obtain a value (in meters) for the astronomical unit?

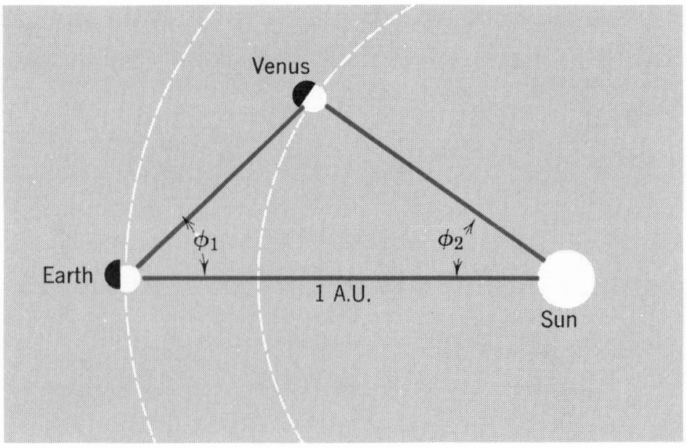

Fig. 1.3 *Triangulation method for the determination of the astronomical unit. Astronomical observations and a knowledge of the period of the Venus orbit are used to obtain the angles ϕ_1 and ϕ_2. Radar-ranging is used to measure the Earth-Venus distance. Then, knowing two angles and the length of one side of the triangle, the base (i.e., the astronomical unit) can be calculated.*

The most precise method available at present for the determination of the astronomical unit involves a combination of angle measurements and *radar-ranging* techniques. Figure 1.3 illustrates the method for the case of the planet Venus. The angles ϕ_1 and ϕ_2 can be determined from astronomical observations and a knowledge of the period of the Venus orbit. Then, the length of the base of the Earth-Venus-Sun triangle (that is, the astronomical unit) can be calculated by trigonometric methods if the length of either of the other two sides is known. A direct measurement of the Earth-Venus distance can be made by radar-ranging, in which a powerful transmitter sends out a burst of radio waves of extremely short duration. These waves travel through space and strike Venus; a small fraction of the reflected waves return to the Earth where they are detected with a sensitive receiver. We know (also by direct measurements) that radio waves travel through space with the speed of light, 3×10^8 m/sec (or 186,000 mi/sec). Therefore, by measuring the time required for the round trip of the waves from the Earth to Venus and back, we have a determination of the Earth-Venus distance. Such experiments have yielded for the astronomical unit the value

$$1 \text{ A. U.} = 1.495\,979 \times 10^{11} \text{ m}$$
$$= 92{,}955{,}700 \text{ mi} \tag{1.4}$$

That is, the Earth-Sun distance is approximately 93 million miles.

The astronomical unit is the standard of length for all distance measurements in the solar system. Indeed, as we will see in the next paragraph, the astronomical unit serves as the length standard even for the measurements of stellar distances.

DISTANCES TO NEARBY STARS

If we wish to determine the distances to *stars,* then we are presented with a problem that requires an entirely different technique because radar-ranging is not possible at stellar distances. Basically, the method used involves the measurement of *parallax,* an effect that is familiar from everyday experience. In Fig. 1.4*a,* if the observer views the object 0 with his right eye, he will see the object at position *A* with respect to the screen. But if he closes his right eye and opens his left eye, he will see the object apparently

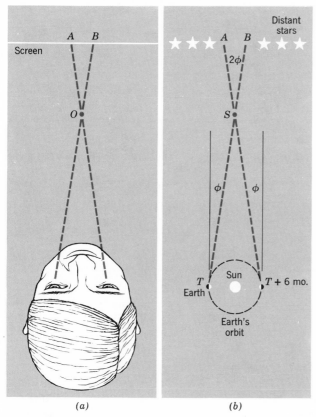

Fig. 1.4 Parallax *causes the apparent shift in position of an object relative to a distant background when the object is viewed from different positions. This method is useful in determining the distances to the nearer stars.*

shift to the position B. Similarly, if a star is sufficiently close to the solar system, then the apparent position of that star on the background of very distant stars will be slightly different depending on the position of the Earth in its orbit. The apparent motion of the star will be greatest if observations are made 6 months apart so that the Earth is on opposite sides of the Sun for the two measurements. The two observation points are therefore separated by a distance of 2 A.U. As shown in Fig. 1.4b, such measurements for the star S define the parallax angle ϕ for the star. The closer that a star lies, the greater will be the angle ϕ. If the parallax angle is 1 second of arc (there are 3600 arc sec in 1 degree), the distance to the star is 2.06×10^5 A.U. or 3.1×10^{16} m.

For the large distances that we encounter in discussing astronomical objects, it is useful to have a length unit even larger than the astronomical unit. Such a unit is the *light year* (L.Y.), the distance that light travels in one year. The speed of light is 3×10^8 m/sec and there are 3.16×10^7 sec in one year. Therefore,

$$
\begin{aligned}
1 \text{ L.Y.} &= (3 \times 10^8 \text{ m/sec}) \times (3.16 \times 10^7 \text{ sec}) \\
&= 9.46 \times 10^{15} \text{ m}
\end{aligned}
$$

Therefore, a parallax of 1 arc sec corresponds to a distance of 3.26 L.Y. In general, if ϕ is measured in arc sec, the distance to a star is given by

$$r = \frac{3.26}{\phi} \text{ L.Y.} \tag{1.5}$$

The number of stars whose distances from us can be measured by the parallax method is limited by the sensitivity with which parallax determinations can be made. About 700 stars are sufficiently close so that parallax measurements have been made to an accuracy of 10 percent or better. The nearest of these stars are *Alpha Centauri* and *Proxima Centauri*, close neighbors in the sky, each of which has a parallax of 0.76 arc sec and is therefore at a distance of $3.26/0.76 = 4.3$ L.Y. Of those stars visible in the Northern Hemisphere, *Sirius* (the brightest star in the sky) is the closest—about 9.5 L.Y.

The parallax method, based on a knowledge of the astronomical unit, is the only direct method available for determining the distances to stars. And this method is useful only for stars out to a distance of approximately 400 L.Y. (corresponding to a parallax of 0.008 arc sec). For more distant stars, we must rely solely on indirect methods that involve the color and the brightness of stars, and, in some cases, the way in which the brightness of a star varies with time (*variable stars*). All such methods are based on determining the average properties of large numbers of stars. All astronomical distance determinations beyond about 400 L.Y. are the result of statistical analyses and are therefore subject to some uncertainty.

GALAXIES

During the 18th century it was discovered that many of the objects in the sky, when viewed with the best telescopes of the day, did not appear as points of light; instead, these objects were *diffuse* and therefore could not be single stars. When the more powerful telescopes of the early 20th century were used to study these objects, it was found that some are glowing clouds of gas or clouds that reflect starlight (these we now refer to as *nebulae*), but that others are clearly composed of large numbers of individual stars. Because the distance determinations available at that time were quite poor, it was not at all clear whether these star groups were part of our local system of stars, the Milky Way, or whether they were separate "island universes." It was not until 1924 that Edwin Hubble of the Mount Wilson Observatory settled the question by analyzing the light from various types of stars in three of the more prominent star groups. Hubble's distance determinations based on these measurements conclusively proved that these objects are not part of the Milky Way and are indeed "island universes." We now refer to these remote cousins of the Milky Way as *galaxies*. Hundreds of thousands of these galaxies have been observed and tabulated, but they are only a tiny fraction of the number that can be observed with our modern telescopes.

When it became clear that the Universe is populated with a large number of galaxies at great distances from the Earth, it was natural to inquire about the distribution of stars in our vicinity. What does our own Galaxy, the Milky

Fig. 1.5 *Cut-away drawing of the 200-in. telescope on Mount Palomar in Southern California. The construction of this telescope, completed in 1948, was first proposed by the astronomer George E. Hale and it is named in his honor. This is the largest optical telescope in the Western Hemisphere, although a larger instrument has recently been constructed in the Soviet Union.*

Way, look like? Would our own Galaxy, if viewed from space, have an appearance similar to those galaxies that *we* can view? Of course, we are at a great disadvantage in making assessments about our own Galaxy because of our position *within* the Galaxy. Large regions of the Galaxy cannot be viewed because of obscuration by local clouds of dust. Therefore, much of the information concerning the overall aspects of our Galaxy has been obtained by indirect methods. The results of these studies have shown that our Galaxy is in no way unique—we live in an ordinary galaxy of ordinary size (diameter $\sim 10^{21}$ m or 10^5 L.Y.). The Milky Way, if viewed from space, would appear very similar to the galaxy NGC 5194 (Fig. 1.6).

The Sun, moreover, does not occupy any special position within our Galaxy. Although it is difficult to pinpoint our location within the Galaxy, it is now believed that the Sun is located in the outer portion of one of the great spiral arms of the Milky Way (see Fig. 1.7). This information confirms our inconspicuousness in the Universe. Not only is the Earth just one of several planets that revolve around an ordinary star, but that star is just one of a hundred billion that comprise an ordinary galaxy.

THE CLUSTERING OF GALAXIES

The study of the distribution of galaxies in space has shown that there are 18 galaxies, of various sizes and shapes, located within 3.5 million L.Y. of our own Galaxy. The clustering of galaxies is the rule, not the exception; almost all galaxies seem to be concentrated in groups. Small clusters may contain a dozen or so galaxies; giant clusters may contain a thousand or more Galaxies. Figure 1.8 shows a cluster of galaxies in the constellation Hercules.

Within a distance of 50 million L.Y. from our Galaxy, thousands of other galaxies have been observed and *billions* are visible through the 200-in. Palomar telescope. Indeed, within the relatively small region of space defined by the "empty" bowl of the Big Dipper, the 200-in. telescope would reveal about a million galaxies!

Fig. 1.6 *The Whirlpool galaxy (NGC 5194) is a spiral galaxy very similar to our own. The small companion galaxy (NGC 5195), located about 14,000 L.Y. away from the larger galaxy, has no clear spiral structure and is classified as an irregular galaxy.*

Fig. 1.7 *The Sun and the solar system are located in one of the great spiral arms of the local system of stars. Because of our location in the galaxy and because the galaxy is relatively flat, we see these stars as a concentrated band in the sky (the Milky Way). The central region of the galaxy is screened from our view by dust clouds.*

THE COMPOSITION OF STARS AND GALAXIES

Of what materials are stars composed? Do galaxies consist of anything more than stars? Modern methods of analyzing the light from stars have shown that almost all stars consist primarily (60 percent or greater by mass) of the simplest element, hydrogen. The next most abundant element in stars is helium and all of the heavier elements constitute no more than a few percent of the total mass.

In addition to stars, all galaxies contain huge clouds of hydrogen. Indeed, new stars are continually being formed by condensation from these clouds. Although a large amount of matter is contained in these galactic clouds, the *density* is extremely low, about 10^8 atoms/m³—far lower than in the best laboratory vacuum. In intergalactic space, the density of gas is even lower, about 1 atom/m³. We live in a Universe that is mostly empty space; even the small amount of matter contained in the Universe is mostly hydrogen!

GALACTIC RED SHIFTS

If a source of light (for example, a star or a galaxy) is moving toward or away from an observer, the character of the observed light is affected. Everyone is familiar with the fact that the tone (or *frequency*) of the sound of a train whistle is higher when the train is approaching than when receding. (As the train passes, the frequency changes and we hear a *whee-oo* sound.)

1.4
MAN'S PLACE
IN THE
UNIVERSE

Fig. 1.8 *A cluster of galaxies in the constellation Hercules.*

This phenomenon of the dependence of frequency on the speed of the source toward or away from the observer is known as the *Doppler effect*. Light behaves in the same way as does sound in this regard.

If a star is moving *toward* us, its characteristic light features are shifted toward the blue, whereas if it is moving *away* from us, the light is shifted toward the red. When the light from distant galaxies is analyzed, a surprising result is obtained. All of these galaxies exhibit a *red shift* in their light, indicating that they are receding from us! Furthermore, the magnitude of the red shift—and, hence, the magnitude of the recessional speed—is greater

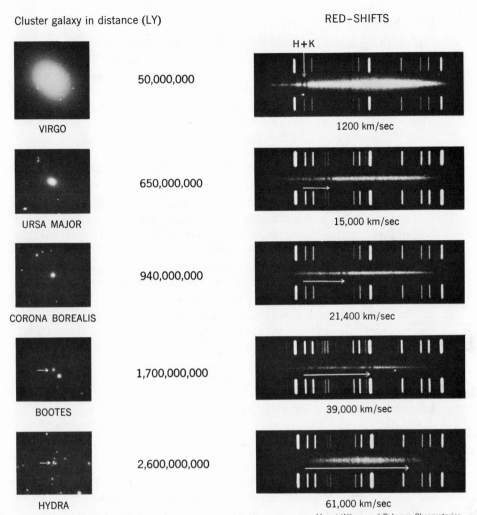

Cluster galaxy in distance (LY) RED-SHIFTS

VIRGO	50,000,000
	H+K 1200 km/sec
URSA MAJOR	650,000,000
	15,000 km/sec
CORONA BOREALIS	940,000,000
	21,400 km/sec
BOOTES	1,700,000,000
	39,000 km/sec
HYDRA	2,600,000,000
	61,000 km/sec

Mount Wilson and Palomar Observatories

Fig. 1.9 (a) *Photographs of individual galaxies in successively more distant clusters of galaxies.* (b) *The Doppler shift (red shift) in the light from these galaxies. The magnitude of the red shift increases with galactic distance.*

for the more distant galaxies (see Fig. 1.9). In fact, there appears to be a direct proportionality between the radial speed S and the galactic distance r, namely,

$$S = Hr \qquad (1.6)$$

where the proportionality constant H is called the *Hubble constant,* in honor of Edwin Hubble who established this relationship in the early 1930s. The value of H is approximately 23 km/sec per million L.Y. (but the uncertainty is large—about 50 percent). That is, a galaxy that is 10^6 L.Y. away recedes from us with a speed of about 23 km/sec and one that is 10^9 L.Y. away recedes with a speed of about 23,000 km/sec. For the most distant galaxies thus far studied ($r \sim 7 \times 10^9$ L.Y.), the red-shift measurements indicate that the recessional speed is more than half the speed of light!

1.4
MAN'S PLACE
IN THE
UNIVERSE

If all the galaxies, in every direction in space, are receding from us, do we not then occupy a special position at the center of the Universe? It is easy to show that this is not the case. Consider the following two-dimensional analogy. Suppose that we blow up a toy balloon that is covered uniformly with spots. As the balloon expands, an observer stationed on one of the spots would see all of the other spots receding from him. Furthermore, the more distant spots would be seen to recede more rapidly than the closer spots. (All distances are measured along the *surface* of the balloon in this analogy.) The same results would be found by observers on each of the other spots. That is, in a uniform expansion, there is an increase of all of the interspot distances. In three-dimensional space, the effects are identical. Consider a box of sand containing several small pebbles. If we imagine the box and its contents to expand in size, an observer stationed on any pebble would observe every other pebble receding from him.

The red-shift measurements are generally interpreted as clear evidence that the Universe is expanding. Since the expansion is apparently uniform, there is no way to specify a "center" of the Universe.

Why is the Universe expanding? Is the Universe finite or infinite? How are stars and galaxies formed? These are some of the questions that we shall discuss when we return to the subject of astrophysics and cosmology in the concluding chapter.

1.5 *The Domains of Physical Theories*

WHEN ARE OUR THEORIES VALID?

In physics we must deal with a variety of quantities, many of which have enormous ranges. *Lengths* vary from the size of elementary particles to the size of the Universe; *times* vary from the lifetimes of transitory particles to the age of the Universe; *masses* vary from the electron mass to galactic masses. How can we incorporate quantities on such vast scales into a theoretical description of the Universe? We have not been (and we may never be) successful in developing a single, all-encompassing theory that describes the variety of phenomena we encounter in microscopic events and in the macroscopic behavior of the matter in the Universe. Instead, we have constructed many different theories, each of which relates to a special area. Consequently, each of these theories is of limited validity.

Newton's famous laws of mechanics, for example, break down when an attempt is made to apply them to cases in which extremely high speeds are involved. Then, we use the *special theory of relativity*. But even this theory cannot be applied when we deal with extremely massive objects or when we attempt to interpret certain phenomena at the enormous distances of the galaxies; for such cases we invoke the *general theory of relativity*. When atomic or nuclear dimensions are involved, Newton's laws give way to *quantum theory* and to *relativistic quantum theory* when both small distances and high speeds are encountered.

How do we know when to use a particular theory? Unfortunately, there is no precise answer to this question. From experience we know that the

PHYSICS AND
THE UNIVERSE
AROUND US

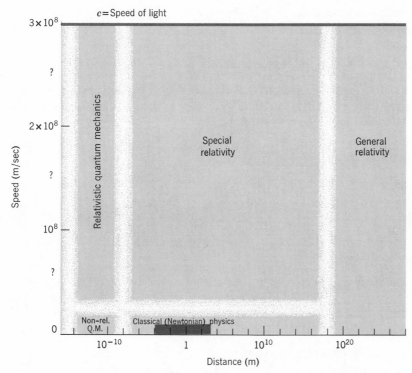

Fig. 1.10 *The distance-speed diagram shows schematically the regions of validity of five of the broadest theories in physics. The theory of* electrodynamics *covers the entire diagram. In the lower left-hand corner is the realm of atoms, molecules, and nuclei—nonrelativistic quantum mechanics. The region of our everyday experience is the small blue line near the middle of the bottom scale.*

relativity theory is certainly necessary when speeds approach the speed of light, and we know that the laws of Newtonian mechanics are adequate for the description of everyday objects if the speeds are very small compared to the speed of light. But at what point must we switch from Newtonian to relativistic mechanics? The answer, of course, depends on the specific case and on the accuracy desired for the particular problem. Nevertheless, we can give some crude guidelines to indicate the regions of validity of some of the more general theories that we currently use.

Figure 1.10 is a *distance-speed* diagram that indicates the areas in which we use five of the broader present-day theories. In addition, the theory of electrodynamics is useful within the entire domain covered by the diagram. There is actually much overlapping in such a diagram and the separations between theory areas are meant to be schematic only.

For distances smaller than the size of a proton (10^{-15} m) and greater than the present estimate of the extent of the Universe (10^{26} m), we have no ideas of what physical theories apply or even whether it makes sense to inquire about the theories for these areas. Therefore, these regions are labeled with question marks.

The scales of distance and speed, which are shown in Fig. 1.10 encompass the total ranges of the quantities about which we have any knowledge in the physical Universe. It is a humbling thought to locate the region of our

1.5
THE DOMAINS
OF PHYSICAL
THEORIES

everyday experience in such a diagram—it is the small solid region near the middle of the bottom scale.

Summary of Important Ideas

Physics is an *experimental* science in which progress toward a deeper understanding of Nature is made by the application of the *scientific method* to empirical facts and observations.

There is no set pattern to the *scientific method*—it is a basic philosophy, colored by personal taste, of the way observation and experiment, reason and logic are used to understand the world around us.

The basic unit of *length* is the *meter* or the *centimeter* (1 m = 100 cm). The *standard* of length is the wavelength of light from krypton-86.

The basic unit of *time* is the *second*. The *standard* of time is the vibration period of cesium-133.

The basic unit of *mass* is the *kilogram* (1 kg = 1000 g). The *standard* of mass cannot yet be stated in terms of atomic quantities with sufficient precision to be generally useful; therefore, the operational standard is a certain block of metal maintained in the international standards depository.

Density is mass per unit volume.

Mass is conserved; that is, mass can neither be created or destroyed—only rearranged. (We shall later see that mass and energy are intimately connected and that the conservation law properly refers to *mass-energy*.)

All physical quantities have *units,* and in equations relating various physical quantities, the *numbers* as well as the *units* on the two sides of the equation must agree.

The fundamental physical units are those of *length, time,* and *mass*.

The basic unit of length for astronomical distance measurements is the distance from the Earth to the Sun (the *astronomical unit*—A.U.).

The only *direct* method of determining the distances to stars is by the measurement of stellar *parallax. All other methods employ statistical techniques* based on the average properties of stars.

The Universe appears to be *expanding*. All distant galaxies are receding from us with speeds that are proportional to their distances.

Distances to the farthest galaxies can be estimated by measuring the Doppler shifts (*red shifts*) in the frequencies of characteristic features of their emitted light.

Questions

1.1

According to Aristotle, if two objects were dropped from a certain height, the heavier object would fall faster. If two objects, one heavier than the other, are tied together with a certain length of string between them, what would Aristotle's "theory" predict for the motion? If the objects were tied together in contact, so that they form essentially a single object, what would Aristotle predict? Can you conclude (as did Galileo) that there is a logical fallacy in Aristotle's "theory"?

1.2

The advancement of science, and physics in particular, has always resulted from the interplay of experiment and theory. Which has been more important in the past, experiment or theory? Which do you believe will play the dominant role in the future? Why?

1.3

List some discoveries in physics that seem to have had important applications in medicine and in communications. (Remember that such items as the invention of the telephone are not "physics discoveries.")

1.4

Discuss whether the following are tenable physical theories:
(a) 40,000 angels can stand on the head of a pin.
(b) The moon is tied to the Earth by a weightless and invisible rope through which any material object will pass without observable effect. (This is why the moon always turns the same face toward the Earth.)

1.5

Why do we construct and then use a certain explanation or *model* of a physical phenomenon when we do not know the fundamental physical basis for the model?

1.6

Look up the definition of *time* in a dictionary. Ignoring those definitions that do not deal with the physical concept of time, comment on the definitions that pertain to time as we use the word in physics. Do these definitions give you a clear understanding of time? Try to devise a better definition of physical time.

1.7

Argue that *all* of the stars that lie at the same distance from the Earth will have the same parallax, regardless of the positions of the stars relative to the plane of the Earth's orbit around the Sun.

1.8

Explain how an artificial Earth satellite could be used to make a precision measurement of the distance from New York to London.

1.9

Do you think that the Universe has always existed or that it had a *beginning*? Explain.

Problems

1.1

The result of a certain experiment gives the following progression of numbers: 1.00, 4.00, 9.00, 16.00, 25.00. Make a hypothesis about the phenomenon under study and predict the next few numbers that should be obtained. Suppose that an experiment is then carried out and the following results obtained: 35.9 ± 0.2, 49.2 ± 0.2, 63.6 ± 0.3, 81.3 ± 0.3. What conclusion would you make about your theory?

1.2

Write out the following numbers in full:
(a) 3.7×10^4 (c) 0.85×10^6
(b) 6.03×10^9 (d) 3×10^{12}

1.3

Write out the following in decimal notation:
(a) 6×10^{-4} (c) 0.39×10^{-5}
(b) 8.6×10^{-7} (d) 3×10^{-12}

1.4

Estimate the following quantities (one significant figure is sufficient) and express them as powers of 10:
(a) The ratio of your height to the diameter of your index finger.
(b) The ratio of your top running speed to the speed of a jet airliner.
(c) The ratio of the height of the Washington Monument to the diameter of a pin head.

1.5

A certain athlete runs the 100-yd dash in 9.4 sec. What would be his expected time for the 100-m dash?

1.6

A certain electronic computer can perform 350,000 arithmetical operations per second. How many microseconds are required (on the average) for each operation?

1.7

Express 100 km/hr in mi/min.

1.8

A certain watch is claimed by the manufacturer to have an accuracy of 99.995 percent. By how many minutes might such a watch be in error after running a month?

1.9

PHYSICS AND THE UNIVERSE AROUND US

The human heart is a marvelous machine. On the average, a man's heart beats at a rate of 72 beats per min. At the age of 70 yr, approximately how many times will his heart have beaten?

1.10

Express 1 microcentury in more conventional units.

1.11

What is the conversion factor between square inches and square centimeters? What is the conversion factor between cubic feet and cubic meters?

1.12

The mass of the Andromeda galaxy (also known by its catalogue designation of M31) is approximately 8×10^{41} kg. Estimate how many stars there are in M31. (Use the fact that the Sun is a rather typical star.)

1.13

The great galaxy in Andromeda is approximately 2.2×10^6 L.Y. away. Express this distance in m.

1.14

The star *Procyon* has a parallax of 0.29 arc sec. What is the distance to *Procyon?*

1.15

A certain cluster of galaxies is found to be receding from us with a speed of 80,000 km/sec. Express the distance to this cluster in light years.

1.16

Assume that the galaxies have always had their present recessional speeds. At what time in the past did the Universe have "zero" size? That is, use the presently accepted value of the Hubble constant to estimate the age of the Universe.

CHAPTER 2
MOTION

If there is one basic theme that extends throughout all areas of physics it is that of *motion*. Atoms in all forms of matter are continually in motion; the motion of electrons produces electrical current; the planets move around the Sun; and even the gigantic galaxies move through space. Historically, the study of simple motions constituted the first extensive application of the scientific method to a problem of the real physical world. Through a series of ingenious experiments and well-constructed, logical arguments, Galileo Galilei (1564–1642) correctly formulated the laws governing the motion of falling bodies; he was the first to explain in detail the motion of projectiles.

2.1 *Average Speed*

THE RATE OF MOVEMENT

The concept of *speed* or *velocity*[1] is familiar to everyone; it is the rate at which something moves. If a certain automobile trip of 30 mi is to be completed in 1 hour, the speed required is, clearly, 30 mi/hr. Two points must be noted about such a statement. First, no mention is made of the *direction* of travel; the trip could be made in a straight line, along a curving highway, or we could just go around the block a sufficient number of times to total 30 mi traveled. Later, we shall be concerned with the *direction* of motion, but now we consider only the simple case of *straight-line motion*. (We allow the possibility of moving in *either* direction along a given straight line.) Second, no mention is made of whether a constant speed of 30 mi/hr is maintained or whether the trip is made by stop-and-go driving. That is, the statement specifies only the *average* speed:

$$\text{Average speed} = \frac{\text{total distance traveled}}{\text{total elapsed time}} \qquad (2.1)$$

---

[1]For the moment, we will use *speed* and *velocity* interchangeably, but in Section 2.5 we will make a distinction between the two terms.

Fig. 2.1 *Galileo.*

Fine Arts Gallery of San Diego, California

We denote velocity by the symbol v and *average speed* or *average velocity* by $\bar{v}$. Also, x stands for the total distance traveled and t for elapsed time. Therefore, in symbols, Eq. 2.1 becomes

$$\bar{v} = \frac{x}{t} \tag{2.2}$$

In general, for any small interval of distance Δx that is traversed in the time interval Δt the average speed is given by[2]

$$\bar{v} = \frac{\Delta x}{\Delta t} \tag{2.3}$$

THE GEOMETRICAL REPRESENTATION OF VELOCITY

If we know that an object which moves in a straight line was at a point labeled by x_0 at a time t_0 and at a point x_1 at a later time t_1 (Fig. 2.2a), we can indicate these facts in a graph of position *versus* time, as in Fig. 2.2b. The net distance traveled is $\Delta x = x_1 - x_0$ and the time interval during which the motion occurred is $\Delta t = t_1 - t_0$. Therefore, according to Eq. 2.3, the average velocity for the motion is the increment of distance moved divided by the corresponding increment of time:

$$\bar{v} = \frac{\Delta x}{\Delta t} = \frac{x_1 - x_0}{t_1 - t_0} \tag{2.4}$$

If the velocity with which an object moves is *constant*, then, clearly, the average velocity is always the same regardless of the particular time interval

---

[2]The symbol Δx does *not* imply the product of Δ and x, but means "a small interval of x" or "an increment of x." In general, a Greek delta, Δ, in front of a quantity means an *increment* of that quantity (see Fig. 2.2).

MOTION

32

Table **2.1** *Some Typical Speeds*

Growth of hair (human head)	5×10^{-9} m/sec
Rapidly moving glacier	3×10^{-6} m/sec = 25 cm/day
Tip of sweep-second hand on wrist watch	10^{-3} m/sec = 1 mm/sec
Running man	10 m/sec
Racing car	70 m/sec = 250 km/hr
Sound in air	330 m/sec
X-15 rocket plane	2×10^{3} m/sec = 2 km/sec
Earth in orbit	3.0×10^{4} m/sec = 30 km/sec
Electron in hydrogen atom	2.2×10^{6} m/sec
Light in vacuum	3×10^{8} m/sec

that is chosen for the computation. An object that moves with a constant velocity of 1 m/sec will have traveled 1 m after 1 sec, 5 m after 5 sec, and 20 m after 20 sec. The distance traveled is therefore a *linear function* of the time; that is, the graph of distance traveled *versus* time is a *straight line*.

Figure 2.3 shows three different distance-time graphs, all of which are

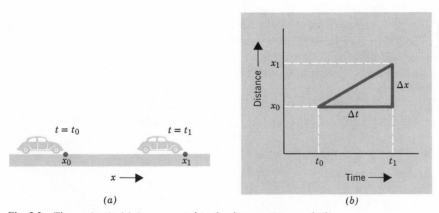

Fig. 2.2 *The motion in (a) is represented in the distance time graph (b).*

straight lines; each corresponds to a different average velocity. As the steepness of the line increases, the average velocity becomes greater. This is a general result: *the slope of a distance-time graph determines the average velocity.*

In discussing straight-line motion we must distinguish between the two possible directions of motion (for example, *right* or *left*, *up* or *down*). We do this by arbitrarily specifying one direction (say, *right*) as *positive* and the opposite direction (*left*) as *negative*. That is, we establish some position as the *origin* and measure all values of x to the right as *positive* and all

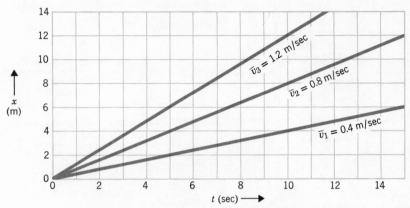

Fig. 2.3 *The greater the slope of the distance-time graph, the greater is the average velocity.*

values of x to the left as *negative*. Actually, it does not matter which direction we choose as positive (x values to the left could be called positive); the only requirement is that we make a choice and use it consistently throughout the discussion of a problem.

2.2 *Instantaneous Velocity*

THE LIMITING AVERAGE VELOCITY

If the motion of an object does not take place with constant velocity, then, in general, the average velocity depends on the particular time interval chosen for the calculation. It seems clear that it would be advantageous to have a method of specifying velocity that gives a unique answer without the necessity of always stating the time interval involved. For this purpose we need the concept of *instantaneous velocity*.

Figure 2.4 shows a distance-time graph that is *curved*. Start with point A, which corresponds to the displacement x_0 at the time t_0. If we take for the final position the displacement x_2 which occurs at time t_2 (point C), we have for the average velocity in this interval,

$$\bar{v}_{02} = \frac{x_2 - x_0}{t_2 - t_0}$$

If we next reduce the time interval to $t_1 - t_0$ we find an average velocity

$$\bar{v}_{01} = \frac{x_1 - x_0}{t_1 - t_0}$$

The slope of the line AB is *greater* than the slope of the line AC; thus, $\bar{v}_{01}$ is *greater* than $\bar{v}_{02}$. If we continue to reduce the time interval (always starting at x_0), the average velocity will increase further.[3] We could, in fact, continue this process indefinitely; we could take smaller time intervals and obtain

---

[3] This *increase of* $\bar{v}$ as Δt is decreased is due to the fact that the distance-time graph has the particular curvature shown. What will happen to $\bar{v}$ as Δt is decreased if the curvature of the distance-time graph is *upward*?

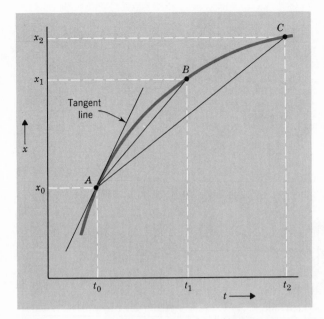

Fig. 2.4 *The average velocity between the initial point A and the final points C and B is greater for shorter time intervals. The* instantaneous velocity *at A results when the time interval is infinitesimally small and is equal to the slope of the tangent line at that point. For this case, note that* $v > \bar{v}$ *for* $\bar{v}$ *calculated with any finite time interval starting at* t_0.

greater and greater average velocities. But if the average velocity for a very small time interval is calculated and this interval is decreased still further, very little change will be produced in the average velocity. We can therefore imagine a time interval so small that any further reduction will not alter the average velocity. This limiting average velocity we call the *instantaneous velocity, v*. Mathematically, we express this result as

$$v = \lim_{t_1 \to t_0} \frac{x_1 - x_0}{t_1 - t_0} = \lim_{\Delta t \to 0} \frac{\Delta x}{\Delta t} \qquad (2.5)$$

This equation states: "The instantaneous velocity v is given by the ratio of Δx to Δt in the limit that Δt approaches zero." Alternatively, "The instantaneous velocity is equal to the average velocity in the limit that the time interval becomes infinitesimally small."

Geometrically, the instantaneous velocity is equal to the slope of the line that is tangent to the distance-time graph at the point in question.

2.3 *Acceleration*

CHANGING VELOCITY

When you "step on the gas" in an automobile, you do so in order to increase the velocity, that is, you *accelerate*. When you apply the brakes, you do so in order to decrease the velocity, that is, you *decelerate* (or undergo a *negative* acceleration). In either case, the essential feature of the motion is that there is a *change* of velocity.

Figure 2.5 shows a case in which the velocity increases linearly (that is, as a straight line) with the time, starting from $v = v_0$ at $t = t_0$. This is therefore a case of *accelerated* motion. Labeling the acceleration with the

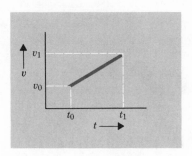

Fig. 2.5 *A case of constant acceleration; the velocity increases linearly with the time.*

symbol a, we can express the average acceleration in a way analogous to that for the average velocity:

$$\bar{a} = \frac{\Delta v}{\Delta t} = \frac{v_1 - v_0}{t_1 - t_0} \tag{2.6}$$

The unit of acceleration must be the unit of velocity divided by the unit of time, in other words, (m/sec)/sec, usually written as m/sec^2.

Just as velocity is the rate of change of distance with time, $\Delta x/\Delta t$, acceleration is the rate of change of velocity with time, $\Delta v/\Delta t$. Compare Eq. 2.4 for $\bar{v}$ with Eq. 2.6 for $\bar{a}$.

UNIFORM ACCELERATION

In Fig. 2.5 the velocity-time graph is a straight line, so that $\bar{a} = a = $ constant. Although there are many physically interesting cases in which the acceleration changes with time, *we shall only consider cases in which the acceleration is constant (or uniform).*

If we displace the velocity-time graph (Fig. 2.5) to the left until the initial time occurs at $t_0 = 0$, the velocity-time graph has the form shown in Fig. 2.6. Then, the acceleration is given by

$$a = \frac{\Delta v}{\Delta t} = \frac{v_1 - v_0}{t_1} \tag{2.7}$$

from which

$$v_1 - v_0 = at_1$$

Because it is the final velocity v_1 in which we are usually interested, we will write this velocity and the corresponding time without a subscript. Therefore, we have

$$v = v_0 + at \tag{2.8}$$

That is, the velocity v at the time t is equal to the initial velocity v_0 plus the additional velocity acquired by virtue of the constant acceleration that acts during the time t.

Next, we seek an expression for the distance traveled in the event that there is a constant acceleration. If we consider the initial position to be at

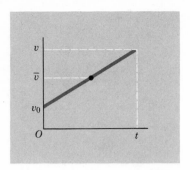

Fig. 2.6 *A velocity-time graph for motion with an initial velocity* v_0. *The average velocity during the time interval from 0 to t is* $\bar{v}$.

the origin of the distance scale so that $x_0 = 0$, the distance traveled is given, as always, by the product of the *average* velocity and the time:

$$x = \bar{v}t \tag{2.9}$$

Referring to Fig. 2.6, it is easy to see that the average velocity, for the case of uniform acceleration, is just the average of v_0 and v. Hence,

$$\bar{v} = \tfrac{1}{2}(v_0 + v)$$

Substitution of this expression for $\bar{v}$ into Eq. 2.9 gives

$$x = \tfrac{1}{2}(v_0 + v)t = \tfrac{1}{2}v_0t + \tfrac{1}{2}vt$$

Using Eq. 2.8 for v, we have

$$x = \tfrac{1}{2}v_0t + \tfrac{1}{2}(v_0 + at)t$$

or, finally,

$$\boxed{x = v_0t + \tfrac{1}{2}at^2} \tag{2.10}$$

That is, the distance traveled is equal to v_0t (the distance that would be traveled in the *absence* of acceleration, as we found previously) plus a term that depends on the acceleration and is proportional to the *square* of the elapsed time.

For the case of *uniform acceleration* and for the initial conditions, $x = 0$ and $v = v_0$ at $t = 0$, we can summarize the results as follows:

Acceleration: $\boxed{\begin{aligned} a &= \text{const.} \\ v &= v_0 + at \\ x &= v_0t + \tfrac{1}{2}at^2 \end{aligned}}$ (2.11)
Velocity:
Distance:

Example **2.1**

In 1970, Don (Big Daddy) Garlits set the world drag racing record by reaching a speed of 240 mi/hr in a distance of $\tfrac{1}{4}$ mile. Assume that Garlits' drag racer accelerated at a constant rate and calculate the acceleration. (The start was from rest.)

We have not derived a formula that allows us to calculate directly the acceleration from a knowledge of the final velocity and the distance traveled.

**2.3
ACCELERA-
TION**

37

We must therefore solve the problem in two steps, starting with a computation of the time required to go $\frac{1}{4}$ mi. We use

$$x = \bar{v}t$$

where $x = \frac{1}{4}$ mi and where the average velocity is

$$\bar{v} = \tfrac{1}{2}(v_0 + v) = \tfrac{1}{2}(0 + 240 \text{ mi/hr}) = 120 \text{ mi/hr}$$

The time required is, therefore,

$$t = \frac{x}{\bar{v}} = \frac{\frac{1}{4}\text{ mi}}{120 \text{ mi/hr}} = \frac{1}{480} \text{ hr} = \frac{1}{480} \text{ hr} \times \frac{3600 \text{ sec}}{1 \text{ hr}} = 7.5 \text{ sec}$$

Since the initial velocity is zero, the acceleration is given by $a = vt$, where v is the final velocity, 240 mi/hr. The result is then

$$a = \frac{v}{t} = \frac{240 \text{ mi/hr}}{7.5 \text{ sec}} = 32 \text{ (mi/hr)/sec}$$

That is, the velocity increases by 32 mi/hr for each sec of travel.

In this case we have given the result in *mixed* units, (mi/hr)/sec, instead of in mi/hr^2 or mi/sec^2 or ft/sec^2, because these mixed units seem to be the easiest to appreciate. There is no sense in establishing arbitrary rules regarding the use of units; we use those units that are most convenient or that convey the best impression of the result. The *physics* is never affected by the choice of units but the comprehension can suffer if a poor choice is made.

Example **2.2**

An automobile traveling at a speed of 30 mi/hr accelerates uniformly to a speed of 60 mi/hr in 10 sec. How far does the automobile travel during the time of acceleration? Using 1 mi = 5280 ft and 1 hr = 3600 sec, we find

$$30 \frac{\text{mi}}{\text{hr}} = 30 \times \frac{(5280 \text{ ft/mi})}{(3600 \text{ sec/hr})} = 44 \text{ ft/sec}$$

Then,

$$a = \frac{\Delta v}{\Delta t} = \frac{88 \text{ ft/sec} - 44 \text{ ft/sec}}{10 \text{ sec}} = 4.4 \text{ ft/sec}^2$$

$$x = v_0 t + \tfrac{1}{2}at^2$$
$$= (44 \text{ ft/sec}) \times (10 \text{ sec}) + \tfrac{1}{2} \times (4.4 \text{ ft/sec}^2) \times (10 \text{ sec})^2$$
$$= 440 \text{ ft} + 220 \text{ ft} = 660 \text{ ft}$$

Suppose next that the automobile, traveling at 60 mi/hr, slows to 20 mi/hr in a period of 20 sec. What was the acceleration?

$$a = \frac{v_1 - v_0}{\Delta t} = \frac{20 \text{ mi/hr} - 60 \text{ mi/hr}}{20 \text{ sec}} = -2 \text{ (mi/hr)/sec}$$

The automobile was *slowing down* during this period so the acceleration is *negative*.

MOTION

2.4 *The Motion of Falling Bodies*

GALILEO'S EXPERIMENTS

When Galileo attacked the problem of the motion of falling bodies, he sought to find a simple relationship connecting quantities that he could measure. By dropping objects of different weights from high places (though probably not from the Tower of Pisa as legend would have it), Galileo quickly concluded that the weight of an object was not a factor in its falling motion. (But see the comments on air resistance in the following section.)

Galileo began his quantitative experiments by rolling balls down inclined planes (see Fig. 2.7). In this way he was able to "dilute" the effect (gravity) that produced the motion of a freely falling body whose motion was too rapid for him to make accurate measurements. Because he lacked a clock to measure the short time intervals involved, he invented a *water clock* for the purpose. He used a large tank from which water was allowed to escape through a small pipe at the bottom. At the start of the motion to be studied, he began collecting the escaping water in a vessel and he removed the vessel from the stream at the end of the motion. By weighing the water collected he could compare the various short time intervals that were involved in his experiments to a precision of about 0.1 sec.

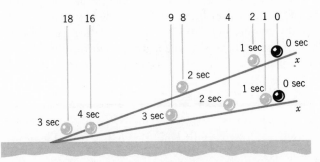

Fig. 2.7 *Galileo measured the times required for a ball to roll various distances down an inclined plane. He found that $x \propto t^2$ for all angles of inclination and thereby verified that the balls were undergoing uniform acceleration.*

Galileo's hypothesis was that a falling object (or one rolling down an inclined plane) would acquire equal increments of velocity in equal intervals of time, that is, the motion would be one of *uniform acceleration*. But he could not test his hypothesis directly because to do so would necessitate measuring the velocity in several very short intervals during the motion. (A moment's reflection will reveal that this is a rather difficult experiment.)

By using a clever geometrical argument, Galileo reasoned that a body undergoing uniformly accelerated motion starting from rest would move, during any interval of time, a distance proportional to the square of that time (compare Eq. 2.10). This conclusion can be tested by simple experiments because it involves measuring *distances* and times instead of *velocities* and times. Using his water clock, Galileo showed that $x \propto t^2$ for balls rolling down his inclined plane. Furthermore, he showed that this relation held for *all* the angles of inclination of the plane for which he could make measurements. By extrapolating his results to an angle of 90°, at which point the plane is no longer involved in the motion, Galileo concluded that a freely falling body obeys the same relation, namely, $x \propto t^2$, and therefore that the body undergoes uniform acceleration when falling freely.

THE ACCELERATION DUE TO GRAVITY

Present-day techniques permit the verification of Galileo's hypothesis regarding falling bodies to be made with high precision. An experiment using a stroboscopic flash is shown in Fig. 2.8. From measurements made with such techniques we find that the acceleration experienced by a body falling freely near the surface of the Earth is approximately 9.8 m/sec², which is equivalent to 32 ft/sec². We give this important number the symbol g:

$$\boxed{\begin{aligned} g &= 9.8 \text{ m/sec}^2 \\ t &= 32 \text{ ft/sec}^2 \end{aligned}} \tag{2.12}$$

A body falling near the surface of the Earth will experience a constant acceleration g only if it falls in a *vacuum*. Usually this is not the case and then the resistive effect of the air comes into play. If the distance of fall is great (for example, a raindrop falling from a cloud), air resistance is an important factor and acts to prevent the object from continuing to accelerate. As a result, the velocity of fall never builds up to a very high value. (What would happen if raindrops were not subject to air resistance?) However, if the distance of fall is not great, air resistance is usually of little significance. In our discussions we consider only the idealized case in which air resistance is negligible.

Example **2.3**

A ball is released from rest at a certain height. What is its velocity after falling 256 feet?

Since the initial velocity is zero, we use Eq. 2.10 with $v_0 = 0$; the distance x is measured vertically downward. Thus,

$$x = \tfrac{1}{2}gt^2$$

Solving for the time of fall, we have

$$t = \sqrt{\frac{2x}{g}}$$

The velocity after falling for a time t is

$$v = gt = g \times \sqrt{\frac{2x}{g}}$$
$$= \sqrt{2gx}$$

Substituting for g and x, we find the velocity to be

$$v = \sqrt{2 \times (32 \text{ ft/sec}^2) \times (256 \text{ ft})}$$
$$= 128 \text{ ft/sec}$$

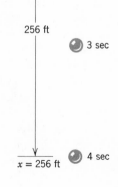

MOTION

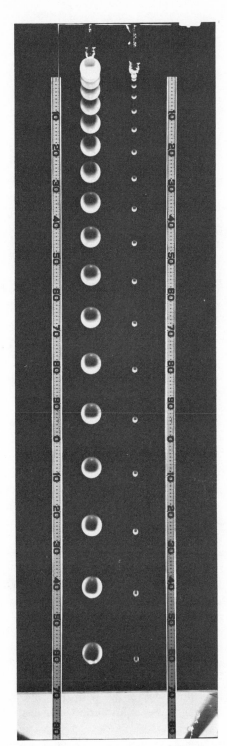

Fig. 2.8 *Two balls of unequal mass fall at the same rate. This is a stroboscopic photograph taken by opening the camera lens and flashing a light source at regular intervals. From the photograph, verify that the balls fall 4 times as far in 16 time units as they fall in 8 time units. The scales are marked in centimeters. Use the fact that g = 9.8 m/sec² and calculate the time interval between successive flashes of the stroboscopic light. (Ans.: approximately 1/40 sec.)*

Fig. 2.9 *Because of air friction, a sky diver will attain a maximum terminal velocity of free fall of approximately 220 mi/hr. When "spread-eagled," the terminal velocity is reduced to about 125 mi/hr.*

In this example we have derived an important and useful result. If an object falls through a height *h*, starting from rest, the final velocity will be

$$v = \sqrt{2gh}$$

2.5 *Vectors*

DIRECTION AND MAGNITUDE

If we wish to describe the motion of an automobile, we could say that the speed is 60 mi/hr. However, this is not a complete specification of the motion; more information is contained in the statement that the velocity is 60 mi/hr in the direction *northeast*. Velocity is a quantity that has both *magnitude* and *direction*. Such quantities are called *vectors*. Another such quantity is *displacement:* an object may move a certain distance but the vector description must include the *direction* of motion as well as the distance traveled. We shall encounter many other examples of vectors in later chapters: force, momentum, electric field, magnetic field, and so on.

Quantities that are completely specified by magnitude alone are called *scalars*. Mass, time, and temperature, for example, are scalar quantities. As we use the terms in physics, *speed* and *velocity* are not identical: speed is a scalar, velocity is a vector. We shall use the term "speed" when we are interested only in the rate at which an object moves and are not concerned with the direction of the motion. When we wish to convey the impression that direction as well as magnitude is important, the term "velocity" will be used.

In order to distinguish vectors from scalars, we will use boldface type for vectors. Thus, the velocity vector will be denoted by **v**. If we are interested only in the *magnitude* of the vector **v**, we will write this as *v*. The magnitude of the *velocity* **v** is just the *speed v*.)

MOTION

In diagrams we represent vectors by arrows. The length of an arrow will be proportional to the magnitude of the vector and the direction of an arrow will be the direction of the vector. We will need only a few simple rules regarding the manipulation of vectors.

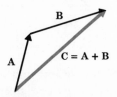

Fig. 2.10 *The addition of vectors **A** and **B** produces the vector **C**.*

In Fig. 2.10 we indicate the rule for vector addition. Graphically, in order to add **A** to **B**, we place the origin of the vector **B** at the head of the vector **A**. Then, the line connecting the origin of **A** with the head of **B** is the sum vector: **C** = **A** + **B**. More than two vectors can be added by simply continuing this procedure, as indicated in Fig. 2.11.

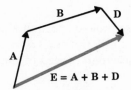

Fig. 2.11 *The addition of three vectors follows the same rule as for the addition of two vectors.*

The negative of a certain vector **A** is another vector, −**A**, which has the same magnitude as **A** but the opposite direction.

The subtraction of one vector from another is the same as adding the negative vector. Figure 2.12 shows the subtraction of **B** from **A** according to two equivalent diagrams. Notice that the vector **B** is placed in different

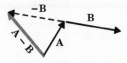

Fig. 2.12 *The subtraction of **B** from **A** according to two equivalent diagrams. Notice that (**A** − **B**) + **B** = **A**.*

positions in the two diagrams, but that is has the same magnitude and same direction in each case. Of course, if one vector is added to another that has the same magnitude but opposite direction, the result is zero: **A** + (−**A**) = 0.

2.5

VECTORS

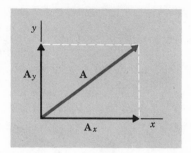

Fig. 2.13 *The vector* **A** *is decomposed into its component vectors,* $\mathbf{A}_x$ *and* $\mathbf{A}_y$: $\mathbf{A} = \mathbf{A}_x + \mathbf{A}_y$.

Just as we can add two vectors and find a single vector that represents the sum, we can also *decompose* a given vector into two vectors the sum of which yields the original vector. This process is particularly useful when the two vectors (called *component* vectors) lie at right angles to one another. We can represent this situation in an *x-y* plot, as in Fig. 2.13. Of course, in such a situation we always have, according to the Pythagorean theorem of plane geometry,

$$A^2 = A_x^2 + A_y^2 \tag{2.13}$$

Example **2.4**

An airplane, whose ground speed in still air is 200 mi/hr, is flying with its nose pointed due north. If there is a cross wind of 50 mi/hr in an easterly direction, what is the ground speed of the airplane?

$$V = \sqrt{v^2{}_{\text{airplane}} + v^2{}_{\text{wind}}}$$
$$v = \sqrt{(200)^2 + (50)^2}$$
$$\quad = \sqrt{42{,}500}$$
$$\quad = 206 \text{ mi/hr}$$

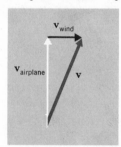

ACCELERATION IS A VECTOR

In addition to displacement and velocity, we also require for the complete description of motion, the vector property of *acceleration*. The *magnitude* of the acceleration of an object is the rate of change of the velocity, and the *direction* of the acceleration is the direction of the *change* in the velocity. If an object is moving in a straight line in the $+x$ direction at the time $t_0 = 0$ with a velocity of 10 m/sec and at a later time $t_1 = 1$ sec has a velocity of 20 m/sec in the same direction, then the *x*-component of velocity has increased and the acceleration vector is in the $+x$ direction: $a = +10$ m/sec². However, if the velocity at t_0 is 20 m/sec and is 10 m/sec at t_1, then the velocity has decreased in the $+x$ direction and therefore the acceleration vector is in the $-x$ direction: $a = -10$ m/sec². Figure 2.14 illustrates this point.

MOTION

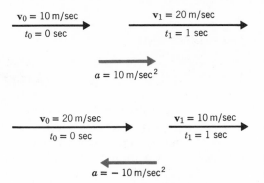

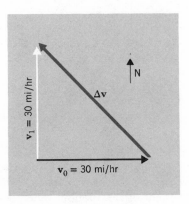

Example **2.5**

At $t_0 = 0$ an automobile is moving eastward with a velocity of 30 mi/hr. At $t_1 = 1$ min the automobile is moving northward at the same velocity. What average acceleration has the automobile experienced?

The figure shows the velocity vectors, $\mathbf{v}_0$ and $\mathbf{v}_1$, and the vector $\Delta\mathbf{v}$ that represents the *change* in velocity. We have

initial velocity + change in velocity = final velocity

That is,

$$\mathbf{v}_0 + \Delta\mathbf{v} = \mathbf{v}_1$$

or,

$$\Delta\mathbf{v} = \mathbf{v}_1 - \mathbf{v}_0$$

The *magnitude* of $\Delta\mathbf{v}$ is (refer to the figure and use the Pythagorean theorem)

$$\Delta v = \sqrt{(30 \text{ mi/hr})^2 + (30 \text{ mi/hr})^2} = \sqrt{1800 \text{ (mi/hr)}^2} = 42.4 \text{ mi/hr}$$

The magnitude of the average acceleration is

$$\bar{a} = \frac{\Delta v}{\Delta t} = \frac{42.4 \text{ mi/hr}}{60 \text{ sec}} = 0.71 \text{ (mi/hr)/sec}$$

The direction of $\Delta\mathbf{v}$, and hence the direction of $\mathbf{a}$ is, from the figure, in the direction *northwest*.

2.5
VECTORS

45

2.6 *Motion in Two Dimensions*

If we drop an object from a certain height, we know that the object will undergo accelerated motion straight downward. What will happen if, just as we drop the object, we also give it some initial velocity (v_{ox}) in the *horizontal* (x) direction? Clearly, the motion will no longer be straight downward but will be at some angle to the vertical (see Fig. 2.15.) Now, we know that velocity is a vector quantity, so we can decompose the velocity vector in this case into a vertical component and a horizontal component. What equations describe the variation with time of these components? If we chose the upward direction as the direction of positive y, the acceleration due to gravity g acts only in the $-y$ direction. Therefore, the y-component of the velocity is

$$v_y = -gt \tag{2.14}$$

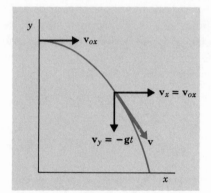

Fig. 2.15 *If an object is dropped and simultaneously given an initial horizontal velocity $\mathbf{v}_{ox}$, this horizontal velocity component remains constant while the vertical component increases linearly with the time. Thus, the motion follows a curved (actually, parabolic) path.*

Since there is no horizontal component of the acceleration, the x-motion is simply

$$v_x = v_{ox} \tag{2.15}$$

These two equations are summarized by the important statement that *the resultant velocity vector consists of two components which act independently.* Only the vertical component of the motion undergoes acceleration, while the horizontal component proceeds at the constant initial velocity v_{ox}.

Figure 2.15 shows the way in which the horizontal and verticaly velocity components combine to give the instantaneous velocity vector **v**. Figure 2.16 is a stroboscopic photograph of two balls that are dropped simultaneously, one with a horizontal velocity component. The picture reveals that the vertical motions are indeed identical.

If we also allow the vertical motion to have an initial velocity v_{oy}, then the equations that describe the motion are:

Acceleration:	$a_x = 0$	$a_y = -g$
Velocity:	$v_x = v_{ox}$	$v_y = v_{oy} - gt$
Displacement:	$x = v_{ox}t$	$y = v_{oy}t - \frac{1}{2}gt^2$

$$\tag{2.16}$$

MOTION

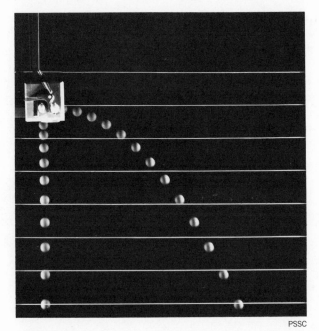

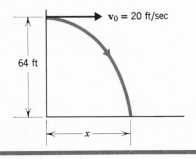

Fig. 2.16 *The two balls were released simultaneously: the one on the left was merely dropped while the other was given an initial horizontal velocity. The vertical components of the motion of both balls are exactly the same. The stroboscopic photograph was taken with a flash interval of 1/30 sec. The distance between the horizontal bars is 15 cm. From the photograph verify that g = 9.8 m/sec². What is the horizontal component of the velocity of the projected ball?*

PSSC

Example **2.6**

An object is released from a height of 64 ft and is given an initial horizontal velocity of 20 ft/sec. When the object strikes the ground, how far has it traveled in the horizontal direction?

Since $v_{oy} = 0$, we have

$y = -\frac{1}{2}gt^2$

The final value of y is -64 ft, so

$-64 = -16t^2$

Thus,

$t = \sqrt{\dfrac{64}{16}} = \sqrt{4} = 2$ sec

Therefore,

$x = v_{ox}t$

$ = (20 \text{ ft/sec}) \times (2 \text{ sec})$

$ = 40$ ft

2.7 *Circular Motion*

ANGULAR VELOCITY

Another very important case of two-dimensional motion is that of motion in a circular path. For simplicity, we assume that the motion is *uniform,* that is, that the *speed* of the object is constant. If it takes 1 sec for the object to make a complete revolution, we say that the object moves at an *angular rate* of 1 rev/sec. But one complete revolution corresponds to 2π radians,[4] so we can alternatively state that the object moves with an *angular velocity* of 2π rad/sec. It is customary to denote angular velocity (measured in radians per second) by the symbol ω.

If we do not have a complete revolution on which to base a calculation, we can still define the angular velocity in a manner entirely analogous to that used for ordinary (or *linear*) velocity. Thus, if an object moves from a point identified by the angle θ_0 to a point identified by the angle θ_1 in a time interval $t_1 - t_0$ (see Fig. 2.17), the *average angular velocity* is

$$\bar{\omega} = \frac{\theta_1 - \theta_0}{t_1 - t_0} = \frac{\Delta\theta}{\Delta t} \tag{2.17}$$

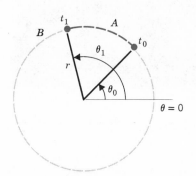

Fig. 2.17 *If the object moves from A to B during the time interval $t_1 - t_0$, the average angular velocity is $(\theta_1 - \theta_0)/(t_1 - t_0)$ rad/sec.*

Since it is only the angular *difference* that is important, the position labeled $\theta = 0$ is arbitrary.

Again, analogous to the case for instantaneous linear velocity, the *instantaneous* angular velocity is defined to be

$$\omega = \lim_{\Delta t \to 0} \frac{\Delta\theta}{\Delta t} \tag{2.18}$$

For uniform circular motion, $\omega = \bar{\omega}$. We shall consider only such cases.

THE PERIOD

The *period* of circular motion is the time required for one complete revolution or cycle of the motion. Clearly, the period and the angular velocity are inversely related since the greater the angular velocity, the shorter the time required to make a revolution. The period is denoted by the symbol τ:

$$\boxed{\tau = \frac{2\pi}{\omega}} \tag{2.19}$$

If an object moves with uniform speed in a circular path with radius r, the distance traveled in 1 period is just the circumference of the circle, $2\pi r$.

[4]See the Appendix for a discussion of radian measure.

The time required for this motion is τ. Therefore, the magnitude of the linear velocity is

$$v = \frac{\text{distance}}{\text{time}} = \frac{2\pi r}{2\pi/\omega}$$

Thus,

$$\boxed{v = r\omega} \qquad (2.20)$$

If the object in Fig. 2.17 is connected to the center of rotation by a rigid shaft, it is important to note that not only does the object move with an angular velocity ω, but every portion of the shaft moves with the *same* angular velocity. Certainly, those portions of the shaft nearer to the center of rotation move with smaller *linear* velocities v, but all parts of the system move with an identical *angular* velocity.

For this case of circular motion, the velocity vector is continually changing in direction (see Fig. 2.18). However, if the motion is uniform, the *magnitude* of the velocity vector (the *speed*) is everywhere the same.

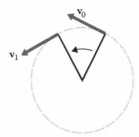

Fig. 2.18 *The velocity vector for circular motion changes direction continually.*

CENTRIPETAL ACCELERATION

If an object moves without acceleration, there is no change of velocity and the velocity vector is therefore constant in magnitude and direction. Similarly, if there is any change of velocity, then there must have been acceleration. This change need not be in magnitude; a change in the *direction* of the velocity vector, even if the magnitude remains constant, requires acceleration. Thus, an object moving uniformly in a circular path is continually accelerated.

We can derive an expression for the acceleration in circular motion by referring to Fig. 2.19. At the time t_0 the velocity vector of the moving object is v_0. At the time t_1 the motion has progressed by an angle $\Delta\theta$ and the velocity vector is v_1. If the motion is uniform, $v_0 = v_1$ even though the directions of v_0 and v_1 are different. The vector Δv represents the change in the velocity during the time interval $\Delta t = t_1 - t_0$.

Notice that the velocity vectors v_0 and v_1 are each perpendicular to the radius lines at A and B, respectively. Therefore, the triangle OAB is similar to the triangle formed by v_0, v_1, and Δv (Fig. 2.19), and we can write

$$\frac{\Delta v}{v} = \frac{\overline{AB}}{r} = \frac{v\,\Delta t}{r} \qquad (2.21)$$

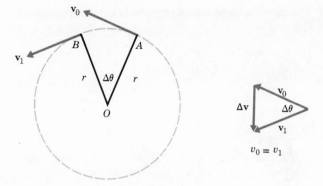

Fig. 2.19 *Velocity diagram of accelerated motion in a circle.*

where v is used for the magnitude of either $\mathbf{v}_0$ or $\mathbf{v}_1$ since these magnitudes are equal, and where $v\,\Delta t$ has been substituted for the distance moved, $\overline{AB}$. Solving for $\Delta v/\Delta t$, we have

$$\frac{\Delta v}{\Delta t} = \frac{v^2}{r} \tag{2.22}$$

But the change in the velocity per unit time is just the *acceleration;* therefore,

$$a_c = \frac{v^2}{r} \tag{2.23}$$

Referring to Fig. 2.19 we see that as Δt becomes small, $\Delta\theta$ also becomes small, and $\mathbf{v}_1$ almost coincides with $\mathbf{v}_0$. Then, $\Delta\mathbf{v}$ is a vector essentially perpendicular to both $\mathbf{v}_0$ and $\mathbf{v}_1$; that is, $\Delta\mathbf{v}$ points toward the center of the circle. The direction of the acceleration vector is the same as the direction of the change in velocity; that is, the acceleration is "center seeking" and is termed *centripetal* acceleration, denoted by a_c.

Using our previous result that $v = r\omega$ (Eq. 2.20), we can also express a_c as

$$a_c = \frac{v^2}{r} = \frac{(r\omega)^2}{r}$$

or,

$$a_c = r\omega^2 \tag{2.24}$$

Example **2.7**

We know that if we drop an object while giving it a horizontal velocity component, the object will fall toward the surface of the Earth with the horizontal velocity remaining constant. With what velocity must an object be projected so that the curvature of its path is just equal to the curvature of the Earth? In such a situation, the object would fall toward the Earth at the same rate that the surface of the Earth curves away from the instantaneous velocity vector; that is, the object would *fall* around the Earth. The height of the object above the surface of the Earth would therefore never decrease and the object would become a satellite of the Earth.

MOTION

50

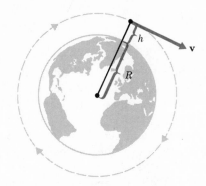

Suppose that we start with the object at a distance h above the surface of the Earth. The radius of the Earth is R so that the radius of the desired circular path of the object is $R + h$.

The centripetal acceleration required to maintain the circular motion is

$$a_c = \frac{v^2}{r} = \frac{v^2}{R + h}$$

This centripetal acceleration is furnished by gravity, so we can substitute g for a_c; thus,

$$g = \frac{v^2}{R + h}$$

As we shall see later when we study gravitation in more detail, the value of g depends on the distance from the center of the Earth. If h is small compared to R (let us take $h = 100$ mi $= 1.6 \times 10^5$ m which is $1/40$ of the Earth's radius), the value of g will be essentially that appropriate for the surface of the Earth, namely, $g = 9.8$ m/sec². Therefore, solving for v, we find

$$
\begin{aligned}
v &= \sqrt{g(R + h)} \\
&= \sqrt{(9.8 \text{ m/sec}^2) \times (6.4 \times 10^6 \text{ m} + 1.6 \times 10^5 \text{ m})} \\
&= \sqrt{(9.8 \text{ m/sec}^2) \times (6.56 \times 10^6 \text{ m})} \\
&= \sqrt{62.7 \times 10^6 \text{ m}^2/\text{sec}^2} \\
&= 7.9 \times 10^3 \text{ m/sec} \\
&= 7.9 \text{ km/sec}
\end{aligned}
$$

or approximately 5 mi/sec.

The period of the motion is

$$
\begin{aligned}
\tau &= \frac{2\pi}{\omega} = \frac{2\pi r}{v} \\
&= \frac{2\pi \times (6.56 \times 10^6 \text{ m})}{7.9 \times 10^3 \text{ m/sec}} \\
&= 5200 \text{ sec} = 87 \text{ min}
\end{aligned}
$$

Therefore, the satellite moving in a circular orbit at a height of 100 mi will require approximately an hour and a half to circle the Earth. Many of the artificial satellites that have been launched during the past few years have orbit characteristics similar to those in this example.

2.8 *The Ultimate Velocity*

According to the formulas we have developed, if an object starts from rest and is accelerated, the velocity after a time t will be $v = at$. Thus, for any nonzero value of a, if we make t sufficiently long, we can have an arbitrarily large velocity v. However, relativity theory tells us that this is not the case. When we deal with velocities that are comparable with the velocity of light, the simple expressions we have developed are no longer adequate. In fact, one of the results of this theory (see Chapter 9) is that there exists a maximum velocity for any physically realizable motion. This ultimate velocity is the velocity of light in a vacuum, $c = 3 \times 10^8$ m/sec. No matter how large an acceleration is impressed on a material object and no matter how long this acceleration is applied, the velocity can never exceed or equal the velocity of light. Thus, the *ultimate* physical velocity is the velocity of light.

Fig. 2.20 *The Stanford Linear Accelerator (SLAC) near Palo Alto, California. This two-mile-long accelerator produces ultra-high velocity electrons for elementary particle research. The velocity of these electrons is only 1.5 cm/sec less than the velocity of light.*

Stanford University

Particles of small mass, such as electrons and protons, can be given extremely high velocities by devices of various sorts. Even the 20-kilovolt electron guns found in television sets produce electrons with velocities of approximately 8×10^7 m/sec or $0.3\,c$. The new Stanford Linear Accelerator (Fig. 2.20) can produce electrons with velocities exceeding $0.9999999999\,c$. But no accelerator can produce a particle with $v = c$ or $v > c$.

Summary of Important Ideas

Speed or *velocity* is the rate at which distance is traveled. *Velocity* is a vector; speed is a scalar.

MOTION

52

On a distance-time graph, the *slope* of the curve is the *speed* or the *velocity*.

Acceleration is the rate of change of velocity. Acceleration is a *vector*.

For objects falling near the surface of the Earth, the horizontal and vertical motions are *independent*. The vertical motion undergoes an acceleration $g = 9.8 \text{ m/sec}^2 = 32 \text{ ft/sec}^2$.

An object moving in a circle has a *centripetal acceleration* that is directed toward the center of the circle.

No object can travel faster than the *speed of light*.

Questions

2.1

Is it possible for a moving body to have **v** and **a** *always* in *opposite* directions?

2.2

The acceleration applied to a certain body is constant in magnitude and direction. Describe situations in which (a) the velocity vector always has the same direction and (b) the body never moves in a straight line.

2.3

In the game of roulette, it is customary to set the ball into motion in the direction opposite to that of the revolving wheel. What is the difference between the motion of the ball relative to the wheel and the motion of the ball relative to the table? Can the ball ever have *zero* instantaneous velocity relative to the table even if the wheel is moving?

2.4

A particle moves uniformly in a circle. Describe the way in which the acceleration vector changes with time.

2.5

Design experiments to measure the following: (a) speed of a pitched baseball; (b) speed of a jet airplane flying at high altitude; and (c) speed of air movement in a light breeze.

Problems

2.1

A driver on an Interstate highway notices the mileage markers at the following times:

Mile	120	11:30 A.M.
	140	11:50 A.M.
	150	12:40 P.M.
	200	1:40 P.M.
	208	1:46 P.M.
	208	1:50 P.M.
	218	2:05 P.M.

(a) Plot a graph of distance *versus* time for the trip.
(b) Indicate on the graph the average speeds for the various straight-line portions of the graph.
(c) What apparently happened near noon and at 1:46 P.M.?

2.2

During a certain automobile trip the speeds at different time intervals were as follows:

1:00 P.M.–2:00 P.M.	$v = 30$ mi/hr
2:00 P.M.–2:30 P.M.	$v = 40$ mi/hr
2:30 P.M.–3:00 P.M.	stopped
3:00 P.M.–4:30 P.M.	$v = 20$ mi/hr
4:30 P.M.–5:00 P.M.	$v = 60$ mi/hr

(a) What was the total length of the trip?
(b) What was the average speed?
(c) How long did it take to go the first 55 mi?
(d) What was the average speed for the first two hours? For the last two hours?

2.3

An automobile moves uniformly a distance of 60 ft in 2 sec; during the next 3 sec it moves only 30 ft. What was the average speed (a) during the first 2 sec, (b) during the next 3 sec, (c) during the 5-sec interval?

2.4

An automobile starts from rest and after 3 sec of uniform acceleration is moving with a speed of 60 ft/sec. (a) What acceleration has the automobile undergone? (b) How far did the automobile travel during the 3-sec interval?

2.5

From a height of 256 ft above the ground, two balls are thrown, one up and one down, each with initial velocities of 20 ft/sec. Compare the velocities when the balls strike the ground. (Try to obtain an answer without performing a calculation.)

2.6

A ball is thrown upward with an initial velocity of 96 ft/sec. How high will it rise? What will be the velocity of the ball when it returns to its original position?

2.7

At time $t = 0$ an object is thrown vertically into the air. It rises to a maximum height of 64 ft and then falls to the bottom of a pit which is 336 ft deep. At what time does the object strike the bottom?

2.8

On the moon the acceleration due to gravity is only $\frac{1}{6}$ as large as on the Earth. An object is given an initial upward velocity of 100 ft/sec at the surface

MOTION

of the moon. How long will it take for the object to reach maximum height? How high above the surface of the moon will the object rise?

2.9

A rocket is launched from the surface of the Earth with an upward acceleration of 4g. After 10 sec, what is the velocity of the rocket and to what height has it risen?

2.10

Two automobiles, each traveling with a speed of 60 mi/hr, crash head-on. The impact velocity is the same as if the cars had been dropped from what height? If it requires $\frac{1}{50}$sec for the cars to come to rest after the initial impact, what acceleration has each experienced?

2.11

The use of *Mach numbers* is one way to specify velocity. Mach 1 corresponds to the speed of sound in air and at sea level is approximately equal to 1100 ft/sec or 730 mi/hr; Mach 2 corresponds to twice the speed of sound, and so forth. A certain high-acceleration rocket can reach Mach 1.2 by the time it has traveled 1000 ft. What is the acceleration? Express the result as a multiple of *g*.

2.12

It requires approximately 2 sec for a parachute to deploy. During this time interval suppose that the speed of the parachutist is decreased uniformly from his free-fall velocity to essentially zero. If the parachutist is falling at the terminal velocity of 200 mi/hr when he opens his parachute, what acceleration does he experience during the deployment? Express the result in terms of *g*. Is parachuting safe for a person who "blacks out" at 5g?

2.13

Each vector in the diagram has a length of 2 units. Graphically determine the following sums:

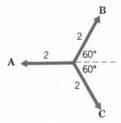

(a) What is the sum **B** + **C**?
(b) What is the sum **B** + **C** + $\frac{1}{2}$**A**?
(c) What is the sum **A** + **B** + **C**?

2.14

At a certain instant an object has the following velocity components: $v_x = 300$ m/sec, $v_y = 400$ m/sec. What is the *speed* of the object at that instant?

2.15

Lay out the following vectors and *graphically* find the sum, first by adding **B** to **A** and then adding **C** to the sum, and next by adding **C** to **B** and then adding **A**.

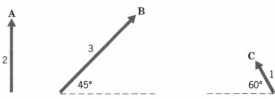

2.16

An object falls through a vertical distance of 256 ft while moving horizontally a distance of 100 ft. What is its horizontal velocity?

2.17

A disc has a radius of 0.2 m and is rotating with an angular velocity of 3 rad/sec. With what speed is a point on the edge of the disc traveling?

2.18

What is the angular velocity and the period of a phonograph turntable operating at $33\frac{1}{3}$ revolutions per minute?

2.19

What is the angular velocity of the Earth's rotation about its axis? What is the speed of a point on the Earth's surface at the equator?

2.20

What is the speed of the Earth in its orbit around the Sun?

2.21

An object is moving in a circular path of radius 9.4 ft and is experiencing a centripetal acceleration of 3 *g*. What is the speed of the object? What is the period of the motion?

CHAPTER 3
FORCE AND MOMENTUM

In the preceding chapter we discussed the topic of *kinematics,* the *geometrical* description of motion. We now extend the discussion to include the *cause* of motion and treat the subject of *dynamics*. In order to make any progress we find that it is necessary to introduce simultaneously *two* new quantities—*force* and *mass*. It is not possible to give precise definitions for force and mass that are based solely on kinematical concepts. Furthermore, the definitions cannot be made independent of one another. For these reasons the logical structure of dynamics is fraught with difficulties for which there appears to be no escape. This fact is a great tribute to the insight of Isaac Newton who was able to construct a workable theory that was capable of explaining in detail many natural phenomena in spite of the basic weakness of its logical foundation.

Because of this logical difficulty, there is no "right way" to proceed with the discussion of dynamics. The approach we elect to use here is not unique, and several other avenues are possible. However, "the proof of the pudding is in the eating," and no matter how we arrive at the Newtonian laws of dynamics, the important point is that these laws correctly describe the motions of macroscopic bodies.

During this century it has been found that the Newtonian laws need modification when exceedingly small distances or extremely high velocities are encountered. The discrepancies between prediction and observation under these circumstances paved the way for the development of the theories of quantum mechanics and of relativity. These limitations of Newtonian theory do not in any way imply that the theory is obsolete or unimportant. Indeed, under almost all ordinary circumstances the Newtonian laws are a correct description of the dynamics of physical systems.

3.1 *The Intuitive Conception of Force*

FORCE AND INERTIA

Force is most commonly appreciated in terms of muscular action. We must exert a "great force" in order to push an automobile but a similar "great force" applied to a large truck produces no motion at all. We know that the truck is much larger than the automobile and contains a greater amount of matter; that is, the truck has a greater *mass* than the automobile. Evidently, the amount of motion that is produced by a given "force" depends on the mass of the body.

The push (or *force*) necessary to start the automobile into motion (even on a level pavement) is rather large, but once in motion it requires much less force to maintain a constant speed. We use the term *inertia* for that property of matter which causes it to resist a *change* in its state of motion. That is, the *inertia* of the automobile must be overcome in order to set it into motion. Similarly, once in motion at constant speed, its inertia prevents the automobile from being stopped unless a sizable force is applied against it (see Fig. 3.1).

Fig. 3.1 *A large force is necessary in order to overcome the inertia of an automobile at rest and set it into motion. Once moving, it requires only a small force (to overcome friction) to maintain motion with a constant velocity.*

We know also that when we apply to an object a force in a certain direction, the object will move in that direction. That is, force is a *vector* quantity.

FRICTION

In addition to any force that we may apply to an object, there will in general be opposing *frictional forces* in operation as well. Even a well-polished ball rolling over a smooth surface will eventually come to rest because of friction. If the ball is not well-polished and if the surface is not smooth, the motion will cease much more quickly. Friction comes about when the irregularities in the surface of an object tend to "snag" on similar irregularities on the surface of another object against which it is sliding or over which it is rolling. Obviously, the less regular the surfaces, the greater the friction will be. Since it is impossible to polish away *all* of the irregularity of any surface (all matter is composed ultimately of "irregular" atoms), friction can never be eliminated, although it can be reduced to exceedingly small values in certain circumstances.

FORCE AND
MOMENTUM

In most of our discussions we shall treat idealized cases in which friction is assumed to be unimportant.

In 1687, Isaac Newton (Fig. 3.2) produced his most important work, the famous *Principia,* in which he presented a new theory of motion. Newton did not "invent" the subject of dynamics; on the contrary, he made the maximum use of previous work; especially the detailed experiments and analyses of Galileo. Newton's great contribution was to synthesize, from his own work and all that went before him, a complete description of the dynamics of bodies in motion. In this section we summarize Newton's results by stating his famous three laws in modern terms, and we briefly discuss the meaning of each law.

National Portrait Gallery

Fig. 3.2 *Sir Isaac Newton (1642–1727), the great English mathematician and physicist who, as a young man, formulated a complete theory of dynamics, discovered the law of universal gravitation, and invented the calculus.*

NEWTON'S FIRST LAW (OR THE LAW OF INTERTIA)

If the net force on an object is zero (i.e., if the vector sum of all forces acting on an object is zero), then the acceleration of the object is zero and the object moves with constant velocity. That is, if an object at rest is subject to no net force, the object remains at rest; if the object is in motion, it maintains a constant velocity. In equation form,

$$\mathbf{F}_{net} = 0 \quad \text{implies that} \quad \mathbf{a} = 0 \quad \text{or} \quad \mathbf{v} = \text{const} \tag{I}$$

This law provides an explanation of the observation, mentioned in the preceding section, that once an automobile is moving, it requires only a small force to maintain constant velocity. If we did not have friction to overcome, the automobile would coast with constant velocity even though *no* force were applied. A hockey puck, sliding over smooth ice, will move a great distance at almost constant velocity even though there is no applied force. Eventually, of course, the puck will come to rest as a result of the nonzero friction that exists between the puck and the ice.

The first law, by itself, gives us only a crude notion regarding force. In fact, we have here only a definition of *zero* force. However, there is the implication that *force* is somehow intimately connected with *acceleration.* The second law states this connection.

NEWTON'S SECOND LAW

The accelerated motion of a body can only be produced by the application of a force to that body. The direction of the acceleration is the same as the direction of the force and the magnitude of the acceleration is proportional to the magnitude of the force. The constant proportionality is the *inertia* or *mass* of the body. In equation form, this law is

$$\boxed{\mathbf{F} = m\mathbf{a}}$$ (II)

Because **a** is a well-defined quantity, this law expresses the relationship between force and mass but it defines neither. It remains for the third law to resolve the situation.

Notice that Eq. II is a *vector* equation. That is, the vectors **F** and **a** are related by the scalar quantity *m*. If the force **F** is in a certain direction, then the acceleration **a** is necessarily in that same direction.

It is essential to realize that whenever we wish to use Eq. II to describe the motion of an object, we must be certain that **F** in this equation is the *total* (or net) *force acting on the object in question*. If there are several individual forces acting on an object, **F** is the *vector sum* of all of these individual forces (Fig. 3.3). Whether the object is exerting forces on other bodies is irrelevant. We need only know the forces that are exerted *on* a body in order to calculate its motion.

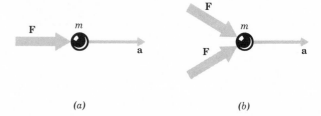

Fig. 3.3 (*a*) *Force and acceleration are in the same direction.* (*b*) *The acceleration is in the direction of the total (or net) force applied to the object.*

Some examples will serve to illustrate these points. (We assume that friction is unimportant in all of these examples.) Figure 3.4 illustrates three basic points: (a) a force is necessary to impart acceleration; (b) the forces applied to an object can cancel so that there is no acceleration; and (c) the net force is always the *vector* sum of the individual forces and may not be in the direction of any one force.

NEWTON'S THIRD LAW

If object 1 exerts a force on object 2, then object 2 exerts an equal force, oppositely directed, on object 1. That is, a force is always paired with an equal reaction force. In equation form,

$$\boxed{\mathbf{F}_{12} = -\mathbf{F}_{21}}$$ (III)

FORCE AND MOMENTUM

where the first subscript denotes the object receiving the force and the second subscript denotes the object exerting the force.

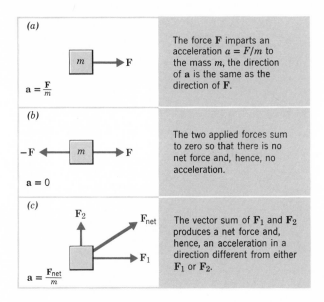

(a) $\mathbf{a} = \frac{\mathbf{F}}{m}$	The force **F** imparts an acceleration $a = F/m$ to the mass m, the direction of **a** is the same as the direction of **F**.
(b) $\mathbf{a} = 0$	The two applied forces sum to zero so that there is no net force and, hence, no acceleration.
(c) $\mathbf{a} = \frac{\mathbf{F}_{net}}{m}$	The vector sum of $\mathbf{F}_1$ and $\mathbf{F}_2$ produces a net force and, hence, an acceleration in a direction different from either $\mathbf{F}_1$ or $\mathbf{F}_2$.

Fig. 3.4 *Three different force situations.*

It is important to understand what the third law means and what it does *not* mean. First, consider a spring balance that is attached to a wall. If we pull on the spring balance it will register the force of the pull; call this force **F** and suppose that it has a magnitude of 10 units. The situation is that shown in Fig. 3.5a. According to Eq. III, the wall should be exerting a force $-\mathbf{F}$ on the spring. How do we know this? We can duplicate the effect of the wall by substituting for it another spring balance and by pulling on this second balance with the force necessary to make the first balance read 10 units while remaining stationary. This new force, **F**′, can be read on the scale and is found to be 10 units (see Fig. 3.5b). The point at which the two springs are joined, or the point at which the spring is attached to the wall (these two equivalent points are labeled A in Fig. 3.5) is not in motion, so the acceleration is zero; thus, the net force must be zero, that is, $\mathbf{F} + \mathbf{F}' = 0$, or $\mathbf{F} = -\mathbf{F}'$.

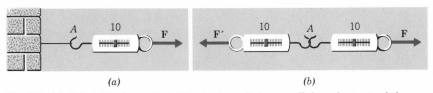

(a) (b)

Fig. 3.5 *(a) A force **F**, of magnitude 10 units, is applied to a wall through a spring balance, (b) To measure the force that the wall exerts on the spring balance, we replace the wall with another balance.*

Next, consider two blocks that are resting on a *frictionless* surface, as in Fig. 3.6. A force **F** is applied to m_1. If m_1 and m_2 are in contact, is this force transmitted to m_2? The third law tells us that whatever force m_1 exerts on m_2, then m_2 must exert this same force on m_1. But the third law does *not* specify the magnitude of this force; only the *second* law can be used to find this quantity. The total mass of the two blocks is $M = m_1 + m_2$ and

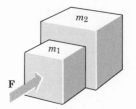

Fig. 3.6 *A force* **F** *is applied to* **m**$_1$. *Is this force transmitted to* **m**$_2$?

the only externally applied force is **F**; therefore, the acceleration of the two blocks together is

$$\mathbf{a} = \frac{\mathbf{F}}{M}$$

The acceleration of m_2 is, of course, also **a**, so the force applied to m_2 that gives rise to this acceleration must be

$$\mathbf{F}_2 = m_2\mathbf{a} = m_2\frac{\mathbf{F}}{M}$$

which is smaller than the external force **F** by the fraction m_2/M. Similarly, the net force acting on m_1 is

$$\mathbf{F}_1 = m_1\mathbf{a} = = m_1\frac{\mathbf{F}}{M}$$

which is just the sum of the externally applied force **F** and the reaction force applied on m_1 by m_2, namely, $-\mathbf{F}_2$:

$$\mathbf{F}_1 = \mathbf{F} - \mathbf{F}_2 = \mathbf{F} - \frac{Fm_2}{M} = \frac{Fm_1}{M}$$

3.3 *The Dimensions of Force*

THE NEWTON

Newton's second law specifies the dimensions of force:

$F = ma$

Therefore, the unit of force is

$$\text{(unit of } mass) \times \text{(unit of } acceleration) = (\text{kg}) \times (\text{m/sec}^2)$$
$$= \text{kg-m/sec}^2$$

If a net force **F**, acting on a mass of 1 kg, causes an acceleration of 1 m/sec^2, the magnitude of the force is 1 kg-m/sec^2. We give the special name *newton* (N) to the unit of force in the metric system:

$$1 \text{ N} = 1 \text{ kg-m/sec}^2 \tag{3.1}$$

Example **3.1**

An object of mass 10 kg is at rest. A net force of 20 N is applied for 10 sec. What is the final velocity? How far will the object have moved in the 10-sec interval?

FORCE AND
MOMENTUM

$$a = \frac{F}{m} = \frac{20 \text{ N}}{10 \text{ kg}} = 2 \text{ m/sec}^2$$

Since the acceleration is constant,

$$v = at = (2 \text{ m/sec}^2) \times (10 \text{ sec}) = 20 \text{ m/sec}$$

$$s = \tfrac{1}{2}at^2 = \tfrac{1}{2}(2 \text{ m/sec}^2) \times (10 \text{ sec})^2 = 100 \text{ m}$$

3.4 The Third Law and the Definition of Mass

MASS RATIOS AND ACCELERATION RATIOS

According to the view we have elected to take of Newtonian dynamics, the first and second laws are essentially only definitions, whereas the third law is indeed a physical *law*. We consider Eq. III to be a real physical law because it allows us, at least in principle, to give a precise definition of *mass*. And, once mass is defined, *force* is defined through Eq. II.

Suppose that we have two objects that are isolated, that is, they interact only between themselves and with nothing else. An example of such a pair of isolated objects would be a star that has a single planet. Even the Sun and Earth are, to a good approximation, an "isolated" pair of objects because for many purposes we can neglect the presence of the moon and other planets. Or we could consider two spacemen in deep space; if Spaceman 1 pushes on Spaceman 2 with a force $\mathbf{F}_{21}$, then 2 pushes on 1 with an equal and opposite force. That is, the third law states

$$\mathbf{F}_{12} = -\mathbf{F}_{21}$$

where $\mathbf{F}_{12}$ means the force *on* 1 *due* to 2 and $\mathbf{F}_{21}$ means the force *on* 2 *due* to 1. Using Eq. II, we can write

$$m_1\mathbf{a}_1 = -m_2\mathbf{a}_2$$

Considering only the magnitudes of the accelerations,

$$m_2 = \left(-\frac{a_1}{a_2}\right)m_1 \tag{3.2}$$

Therefore, if we were to select m_1 as our standard mass we could determine m_2 by measuring the acceleration ratio, a_1/a_2. (The negative sign indicates only that the accelerations are in opposite directions.) Thus, the third law provides a method of uniquely defining *mass;* then, Eq. II gives a similarly precise definition of *force*.

3.5 Frames of Reference

INERTIAL REFERENCE FRAMES

Newton's laws involve the concept of *acceleration*. In order to measure acceleration we must specify some reference marks with respect to which the measurement is made. These reference marks constitute a *frame of reference* in the same way that a set of coordinate axes serves a reference system for the plotting of graphical data. Not all reference frames are equally useful. If we set out to study the dynamics of planetary motion, we would not choose a coordinate system that is fixed with respect to the Earth; in such a system the planets wander in a complicated manner across the sky and undergo apparent motions that are not indicative of the forces actually acting on them. That is, in an Earth-fixed coordinate system, Newton's laws cannot be used directly to describe the motion of planets. However, if we choose a reference frame that is fixed with respect to the distant stars, the motion of the planets is found to conform to Newton's laws. *Any* reference frame in which Newton's laws are a correct description of the dynamics of moving bodies is called an *inertial reference frame.*

To Newton, the distant stars, which appeared to have fixed positions in space, satisfied the need for a basic inertial reference frame. We now know that these stars are not "fixed" with respect to the Earth (or even with respect to the Sun) but undergo continual and complicated motions. The specification of an inertial reference frame is therefore not a simple problem. However, for all but the most sophisticated analyses, we can consider the distant stars to be fixed and to constitute an acceptable inertial reference frame.

AN INFINITY OF INERTIAL REFERENCE FRAMES

The distant stars do not specify the only possible inertial reference frame. We can find many other reference frames in which Newton's laws are also valid. The Earth undergoes a complicated motion against the background of the distant stars. But if we confine our attention to small-scale phenomena that take place over relatively short periods of time, the motion of the Earth will not influence the phenomena to any appreciable extent. Therefore, in many practical situations, Newton's laws will be valid in a coordinate system fixed with respect to the Earth. Indeed, for most everyday applications of Newton's laws we find an Earth-fixed coordinate system to be quite adequate.

Suppose that we set up a laboratory in a large box that is at rest on the Earth. We equip ourselves with suitable meter sticks, clocks, and spring balances so that we can test Newton's laws. By making various measurements we can verify, for example, that $\mathbf{F} = m\mathbf{a}$ is a valid equation. The laboratory coordinate system therefore is an inertial reference frame. Next, someone removes our box laboratory and places it on a train that is moving with constant velocity. (Let us suppose that our box contains no windows so that we cannot measure the velocity of the train.) If we repeat the measurements that were made when the laboratory was at rest, what will we find? We will find exactly the same results! That is, Newton's laws are also valid in the moving reference frame. The reason is easy to see. A measurement of a certain acceleration requires the measurement of a time interval and the difference between two velocities:

$$\mathbf{a} = \frac{\mathbf{v}_1 - \mathbf{v}_0}{t_1 - t_0}$$

If we add a constant velocity **v** to the coordinate system (by transferring it to the moving train), we have for our new velocities, $\mathbf{v}'_1 = \mathbf{v}_1 + \mathbf{v}$ and $\mathbf{v}'_0 = \mathbf{v}_0 + \mathbf{v}$. Therefore, the new acceleration $\mathbf{a}'$ is

$$\mathbf{a}' = \frac{\mathbf{v}'_1 - \mathbf{v}'_0}{t_1 - t_0} = \frac{(\mathbf{v}_1 + \mathbf{v}) - (\mathbf{v}_0 + \mathbf{v})}{t_1 - t_0} = \frac{\mathbf{v}_1 - \mathbf{v}_0}{t_1 - t_0} = \mathbf{a}$$

Thus, the acceleration measured in the moving reference frame is exactly equal to that measured in the reference frame at rest.

We may therefore draw the following important conclusion: *If Newton's laws are valid in a certain reference frame, then they will also be valid in any other reference frame that moves with constant velocity with respect to the first frame* (see Fig. 3.7). Thus, there is no *one* reference frame that is preferred over all others.

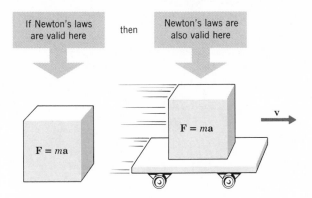

Fig. 3.7 *The addition of a constant velocity does not invalidate Newton's laws.*

3.6 *Mass and Weight*

WEIGHT IS A FORCE

It is important to distinguish between the concepts of mass and weight. *Mass* is an intrinsic property of matter. The mass of a body is a measure of the amount of material in the body. The body contains the same number of atoms regardless of its location, whether on the Earth, on the moon, or in space. Its mass is the same in all of these places, but the *weight* of the body is different. *Weight* is a measure of the gravitational force on a body. A body that has a certain weight at the surface of the Earth will have a different weight on the moon because the gravitational force exerted by the Earth on an object at the Earth's surface is greater (by about a factor of 6) than the gravitational force exerted by the moon on the same object at the moon's surface. In distant space, where the gravitational force due to any other object is negligible, the weight of a body is zero.

In general, the weight w of an object is related to its mass m by

$$w = mg \tag{3.3}$$

where g is the acceleration due to gravity *appropriate for the place at which the measurement is made.* Just as g varies slightly over the surface of the

Earth and is different on the moon or other planets, so also does the weight vary. It must be emphasized that *g* does *not* depend on the mass *m*.

Weight is indeed a *force;* when a body is dropped, it is the *weight* of the body that produces acceleration.

It is often said that an object falling freely is *weightless*. Such a statement, according to our definition of weight, is misleading. At any given height above the surface of the Earth there will be a definite value of the gravitational force on a body; therefore, the body has a definite weight. However, if the object is falling freely and is inside a container that is also falling freely (such as an astronaut in a space capsule), the object will exert no force on the walls of the container—it will float freely and will appear (relative to the container) to be weightless. If a freely falling mass is attached to a spring scale in a freely falling container (Fig. 3.8), the scale will register "zero" weight. But the true weight of the mass is *w* = *mg,* where *g* is the acceleration due to gravity at the position of the mass.

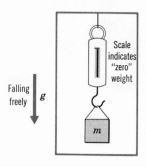

Fig. 3.8 *A freely falling mass in a freely falling container exerts no force on the spring scale—it is apparently "weightless."*

3.7 *Momentum*

FORCE AND CHANGE OF MOMENTUM

In addition to the concepts of force, mass, and acceleration, there are other physical quantities that are of importance in dynamical situations. One of these quantities is *momentum,* which, for a material object, is defined to be the product of the object's *mass* and its *velocity*. Because mass is a scalar and velocity is a vector, the product *m*v is a vector; we give to the momentum vector the symbol **p**:

FORCE AND MOMENTUM

$$\text{Momentum} = \mathbf{p} = m\mathbf{v} \tag{3.4}$$

Even though we have defined and used the concepts of mass and velocity, it proves convenient and useful to introduce the idea of momentum as well. Not only is momentum a useful concept when describing the interactions of the parts of a system, but momentum is even more generally useful. (For example, a photon of light possesses momentum even though it has no mass.) Frequently, we will use the term *linear momentum* for **p** in order to distinguish it from *angular momentum,* which is introduced in the following section.

Newton originally stated his second law, not in terms of (mass × acceleration) as we have indicated, but in terms of the time rate of change of momentum. That is,

$$\overline{F} = \frac{\Delta p}{\Delta t} \quad \text{and} \quad F = \frac{\Delta p}{\Delta t} \tag{3.5}$$

Using the definition of **p** and with the mass constant in time we have

$$F = \frac{\Delta(m\mathbf{v})}{\Delta t} = m \lim_{\Delta t \to 0} \frac{\Delta \mathbf{v}}{\Delta t}$$

or, using the definition of acceleration,

$$F = ma \tag{3.6}$$

so that we recover the form of the second law that we used in Eq. II. *Equation 3.5 is the most general statement regarding force.* In the special case that the mass remains constant, Eq. 3.5 is equivalent to Eq. 3.6.

Example 3.2

A 100-kg man jumps into a swimming pool from a height of 5 m. It takes 0.4 sec for the water to reduce his velocity to zero. What average force did the water exert on the man?

The man's velocity on striking the water was (see Example 2.3)

$$v = \sqrt{2gh} = \sqrt{2 \times (9.8 \text{ m/sec}^2) \times (5 \text{ m})} = 10 \text{ m/sec}$$

Therefore, the man's momentum on striking the water was

$$\mathbf{p}_1 = m\mathbf{v} = (100 \text{ kg}) \times (10 \text{ m/sec}) = 1000 \text{ kg-m/sec}$$

The final momentum was $\mathbf{p}_2 = 0$, so that the average force was

$$\overline{F} = \frac{\Delta p}{\Delta t} = \frac{\mathbf{p}_2 - \mathbf{p}_1}{\Delta t} = \frac{0 - 1000 \text{ kg-m/sec}^2}{0.4 \text{ sec}} = -2500 \text{ N}$$

The negative sign means that the retarding force was directed opposite to the downward velocity of the man.

CONSERVATION OF LINEAR MOMENTUM

Next, we consider the importance of the concept of linear momentum. Suppose we have a system that is *isolated;* that is, the constituents of the system interact with one another but there is no outside agency that acts on them in any way. Truly isolated objects are not possible in the real physical world, but a group of objects whose mutual interaction is much greater than their interaction with other objects can frequently be treated as if they are isolated. For example, consider two hockey pucks that slide (almost) without friction over ice and collide with one another. The individual motions are influenced in a small way by the gravitational forces due to surrounding objects and to a greater extent by friction, but the collisional interaction between the two pucks so overwhelms these other interactions that for most purposes the latter can be neglected.

If a group of objects constitutes an isolated system, there is no net force on this system. Since $\mathbf{F} = 0$, Eq. 3.5 requires that there be no change of linear momentum with time; in other words, $\mathbf{p} = \text{const}$. The momentum $\mathbf{p}$ is the *total* linear momentum of the system and is equal to the vector sum of the individual momentum vectors for all of the particles in the system. This is an extremely important result:

If there is no external force applied to a system, then the total linear momentum of that system remains constant in time.

This is the statement of the *principle* (or *law*) *of the conservation of linear momentum.* The principle is valid not only in classical systems but in quantum mechanical situations as well.

The momentum to which the conservation principle applies is the *total linear momentum* of the system. If the system consists of two objects, m_1 and m_2, then

$$\mathbf{p} = \mathbf{p}_1 + \mathbf{p}_2$$
$$= m_1\mathbf{v}_1 + m_2\mathbf{v}_2 = \text{const.} \qquad (3.7)$$

Suppose we have two masses at rest that are connected by a compressed spring, as in Fig. 3.9. (We will neglect the mass of the spring compared to

Fig. 3.9 *The motion of the masses after release is determined by the conservation of linear momentum.*

m_1 and m_2). In this condition the total linear momentum of the system is zero and the conservation law tells us that it must always be zero. After the release of the spring, the objects move away from each other with velocities $\mathbf{v}_1$ and $\mathbf{v}_2$. Thus,

$$\underbrace{0}_{\substack{\text{before} \\ \text{release}}} = \underbrace{\mathbf{p}_1 + \mathbf{p}_2}_{\substack{\text{after} \\ \text{release}}}$$

so that

$$m_1\mathbf{v}_1 + m_2\mathbf{v}_2 = 0 \qquad (3.8)$$

FORCE AND
MOMENTUM

68

or, considering only the magnitudes of the velocities,

$$m_1 v_1 = -m_2 v_2 \qquad\qquad (3.9)$$

where the negative sign indicates that the velocities are oppositely directed.

Example **3.3**

A high-powered rifle whose mass is 5 kg fires a 15-g bullet with a muzzle velocity of 300 m/sec. What is the recoil velocity of the rifle?

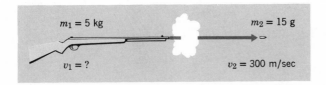

From the conservation of momentum we have

$$m_1 \mathbf{v}_1 = -m_2 \mathbf{v}_2$$
$$\mathbf{v}_1 = -\frac{m_2 \mathbf{v}_2}{m_1} = -\frac{(15 \times 10^{-3}\,\text{kg}) \times (300\,\text{m/sec})}{5\,\text{kg}} = -0.9\,\text{m/sec}$$

This is a sizable recoil velocity and if the rifle is not held firmly against the shoulder, the shooter will receive a substantial "kick." However, if he *does* hold the rifle firmly against his shoulder, the shooter's body as a whole absorbs the momentum. That is, we must use for m_1 the mass of the rifle *plus* the mass of the shooter. If his mass is 100 kg, then the recoil velocity (now of the rifle plus shooter) is

$$v_1 = -\frac{(15 \times 10^{-3}\,\text{kg}) \times (300\,\text{m/sec})}{5\,\text{kg} + 100\,\text{kg}}$$
$$\cong -0.045\,\text{m/sec}$$

This magnitude of recoil is quite tolerable.

Example **3.4**

In the preceding example, we needed the value of the muzzle velocity of the bullet in order to calculate the recoil velocity. Momentum conservation can also be used to measure such quantities. Suppose we fire the 15-g bullet into a 10-kg wooden block that is mounted on wheels and measure the time required for the block to travel a distance of 0.45 m. This can easily be accomplished with a pair of photocells and an electronic clock. If the measured time is 1 sec, what is the muzzle velocity of the bullet?

The recoil velocity of the block is 0.45 m/sec, and from momentum conservation we have

$$m_1 \mathbf{v}_1 = m_2 \mathbf{v}_2$$

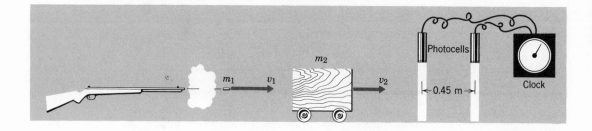

(Here, we do not have a negative sign because both velocities are in the same direction. Also, we take m_2 to be 10 kg, that is, we neglect the mass of the bullet embedded in the block.) Then,

$$v_1 = \frac{m_2 v_2}{m_1} = \frac{(10 \text{ kg}) \times (0.45 \text{ m/sec})}{15 \times 10^{-3} \text{ kg}}$$
$$= 300 \text{ m/sec} \cong 985 \text{ ft/sec}$$

In this example the bullet comes to rest in the block and imparts its momentum to the block. The process by which the bullet stops is a complicated one, but we need to know none of the details in order to calculate the velocity by using momentum conservation. This is indeed a powerful physical principle!

MOMENTUM CONSERVATION AND THE THIRD LAW

Previously, we used Newton's third law to express, in Eq. 3.2 the ratio of two isolated and interacting masses in terms of the ratio of their accelerations. Thus, the third law permits a definition of mass in terms of kinematical quantities. Equation 3.9, written in the form

$$\frac{m_1}{m_2} = -\frac{v_2}{v_1} \tag{3.10}$$

states that we can use momentum conservation to determine the ratio of two masses by measuring the ratio of their velocities. But we obtained the momentum conservation law from Eq. 3.5 which is a statement of Newton's *second* law. Can we therefore avoid the necessity of introducing the third law, since it apparently serves no function other than to define mass, and obtain all of the machinery of dynamics from the second law?

The answer to this question is an emphatic "no." The reason is that we have really used the third law in obtaining the momentum conservation law. Consider the statement that led directly to momentum conservation: *If a group of objects constitutes an isolated system, there is no net force on this system*. This statement can be true only if the third law is obeyed, for if object 1 exerts a force on object 2 there will be a net force on the system unless object 2 exerts an equal and opposite force on object 1. An isolated system cannot accelerate itself; this is guaranteed by the third law, in the guise of momentum conservation.

FORCE AND
MOMENTUM

3.8 *Angular Momentum*

A NEW CONSERVATION PRINCIPLE

From Newton's laws we have obtained the important result that if the net force on an object is *zero*, the linear momentum will remain constant. There is a similar conservation principle that applies to *rotational* motion. Suppose that we have a top spinning on some surface. Friction acts between the tip and the surface, and eventually the top slows down and topples over. But if we imagine a perfectly smooth surface so that there is no friction present, the top will continue to spin indefinitely with the same angular velocity. Or, if we set into motion a ball attached by a string to a pivot point (Fig. 3.10), the ball will never cease its rotation unless some force, such as friction, acts on it. We can then state, that in the absence of any external influence, the *state of rotation* of an object will remain constant.

Fig. 3.10 *In the absence of friction, the ball will maintain a constant state of rotation around the pivot point.*

In order to give the idea more substance, we define a new quantity called *angular momentum* which is the product of an object's linear momentum $m\mathbf{v}$ and the distance r of the object from the rotation point. In Fig. 3.11, the angular momentum of m around the point O is

$$L = mvr \qquad (3.11)$$

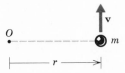

Fig. 3.11 *The angular momentum of m around O is $L = mvr$.*

This quanity, $L = mvr$, will remain constant if no external agency acts to change the state of rotation of m. This is the principle of *angular momentum conservation*.

Because the *linear* velocity v is equal to the product of r and the angular velocity ω (see Eq. 2.20), we can use $v = r\omega$ to express the angular momentum as

$$L = mr^2\omega \qquad (3.12)$$

Fig. 3.12 *By drawing her arms to her sides, the revolving ice skater decreases the amount of mass at large distances from the rotation axis. Because angular momentum must be conserved, this causes the skater's angular velocity to increase.*

Example **3.5**

A satellite of mass m moves around the Earth as shown (actually, the path is an *ellipse*). Which instantaneous velocity is greater, v_1 (at point P) or v_2 (at point A)?

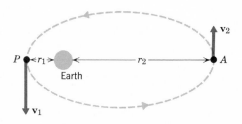

Considering the Earth as a fixed object and neglecting the influence of the Sun and other planets, the angular momentum of the satellite around the Earth is constant. Therefore,

$$mv_1r_1 = mv_2r_2$$

Since $r_1 < r_2$, we must then have

$$v_1 > v_2$$

The velocity is greatest when the satellite is nearest the Earth; this point is called the *perigee* (labeled P in the diagram). The velocity is least at the farthest point from the Earth—the *apogee* (A) of the orbit.

THE DIRECTION OF ANGULAR MOMENTUM

Angular momentum is actually a *vector* quantity. The *magnitude* of the angular momentum vector has already been defined, but the *direction* of the vector must be specified separately. If a particle is moving in a circular path around a certain point, the angular momentum vector is defined to have the direction in which a right-hand screw would advance if moved in the same sense (see Fig. 3.13). Thus, for motion in a plane, the angular momentum vector is *perpendicular* to that plane. In an x-, y-, z-coordinate system, a particle that moves in the x-y plane around the origin has an angular momentum vector that lies along the z axis. Alternatively, if the fingers of the right hand are curled in the direction of the motion of the particle, the thumb will point in the direction of **L**.

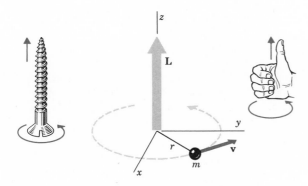

Fig. 3.13 *The direction of the angular momentum vector is the same as the direction of advance of a right-hand screw moving in the same sense; or, if the fingers of the right hand are curled in the direction of motion, the thumb points in the direction of* **L**.

Summary of Important Ideas

Except for objects moving with velocities approaching the velocity of light and for effects that take place on an atomic scale, *Newton's laws* are essentially a correct description of the dynamics of physical systems.

If Newton's laws are valid in a certain frame of reference, they are also valid in *any* frame that moves with *constant linear velocity* relative to the first frame.

A body can be accelerated only if a *force* is applied to it. Force is the time rate of change of momentum.

The motion of a body is determined by the net force acting *on* the body and does not depend on the forces that the body exerts on other bodies.

Mass is an intrinsic property of matter; *weight* is the *gravitational force* on an object.

Force, linear momentum, and angular momentum are all *vectors*.

If no external force is applied to a system, the linear momentum of the system remains *constant* in time.

The angular momentum of a system remains *constant* in time if no external force acts to change the state of rotation.

Questions

3.1

Explain how it is possible for an experienced fisherman to land a 100-lb fish with a line rated at 10-lb breaking strength. Does the fisherman ever use only the line to pull the fish out of water and into the boat?

3.2

A heavy weight is suspended from the ceiling by a length of string. Hanging downward from the weight is another length of the same string. Which string will break when (a) a steady pull is exerted on the lower string and (b) a sudden pull is exerted on the lower string? Explain.

3.3

List the controls in an automobile which, when activated, tend to make a passenger alter his position in his seat. (There are at least four.) What is the nature of the acceleration produced by each?

3.4

A rocket is coasting at constant velocity in free space. When the rocket engine is turned on, the rocket begins to accelerate. Explain how this is possible since there is no matter in the vicinity that can exert a force on the rocket.

3.5

What is the principle of operation of a *recoilless* rifle?

3.6

A wheel can turn freely in a horizontal plane on a vertical axle that is stuck in the ground. A cat is sitting on the rim of the wheel.
(a) The wheel is not in motion. Suddenly the cat begins to walk around the rim. What happens to the wheel and why?
(b) The wheel is spinning at a constant rate with the cat stationary at one point on the rim. Suddenly the cat begins to move toward the center of the wheel. What happens to the wheel and why?

3.7

Consider a stool top that can move freely about a pivot at the center. A boy sits on the stool and holds a spinning top with its axis vertical; the stool is at rest. Suddenly, the boy grasps the top and stops the spinning. What is the result?

3.8

Would it be possible to design a practical helicopter with only one set of rotating blades? Explain.

Problems

3.1

A 10-kg object is pulled over a smooth surface by a 20-N force. What is the acceleration?

3.2

What is the weight of a 100-kg man at the surface of the Earth?

3.3

Two forces, one twice as large as the other, both pull on an object in the same direction ($m = 100$ g) and impart to it an acceleration $a = 3$ m/sec. What are the magnitudes of the forces? If the smaller force is removed, what is the new acceleration?

3.4

A 40-kg block is sliding over a rough surface. At a certain instant the velocity of the block is 2 m/sec. If the block is slowed to rest by friction in a uniform manner after the block has traveled 40 m, what was the average frictional force?

3.5

When a certain type of bullet ($m = 10$ g) is fired from a rifle, a force of 3000 N is exerted for a millisecond. What is the muzzle velocity of the bullet?

3.6

A 50-kg boy is on a 10-kg sled which is at rest on smooth ice. The boy begins to throw 1-kg snowballs horizontally and in the same direction with a velocity of 40 m/sec and at a rate of one per second. Neglect the mass of the snowballs compared to the mass of the boy and the sled and calculate the average force exerted on the sled. How rapidly will the sled accelerate?

3.7

A ball ($m = 100$ g) is thrown against a wall with a velocity of 10 m/sec and it rebounds with a velocity of the same magnitude. If the ball was in contact with the wall for 1 msec (10^{-3} sec), what was the average force exerted on the ball by the wall? (Find the magnitude and the direction of the force.)

3.8

In Example 3.4 it is found that the bullet penetrates 10 cm of wood. Assume that there is a uniform decrease of the velocity to zero. How long did it take for the bullet to come to rest? What average force did the wood exert on the bullet?

3.9

In "shoot-'em-up" movies, the villain is often knocked down by a single bullet from the hero's gun. A .357 magnum bullet has a mass of approximately 10 g and a muzzle velocity of 400 m/sec. What would be the recoil velocity of an average-size man struck by such a bullet? Is it likely that he would be knocked down?

3.10

A cannon of mass 1000 kg is mounted on wheels and rests on a flat surface. If a shell of mass 10 kg is fired horizontally with a muzzle velocity of 200 m/sec, what will be the recoil velocity of the cannon? If the cannon were attached rigidly to the deck of a ship of mass 20×10^6 kg which is moving with a velocity of 10 km/hr, what will be the velocity of the ship after the cannon is fired *forward?*

3.11

A 10-kg cart is moving over a horizontal surface at the constant velocity of 2 m/sec. A 2-kg lump of clay is dropped into the cart from a height of 4 m and it sticks to the bed. Describe the subsequent motion. What happened to the vertical momentum?

3.12

A 0.25-kg block is whirled in a circle at the end of a 2-m rope with an angular velocity of 2 rad/sec. What is the angular momentum of the block?

3.13

In the previous problem, suppose that a 0.1-kg chunk of the block suddenly breaks loose. Describe the subsequent motion of both pieces of the block.

CHAPTER 4
THE BASIC FORCES
OF NATURE

In our everyday experience we encounter a variety of forces—the muscular force exerted to open a door, the frictional force in the door hinges, and the elastic force in the door spring; the force that the atmosphere exerts on a barometer and the force that the Earth exerts on the moon; the electrical force that starts an automobile engine, the hydraulic force that operates the brakes, or the mechanical force that stops the car if we are unfortunate enough to collide with a lamp post. In spite of the large number of names that we have given to forces that we use or must overcome, there are only *two* basic forces that govern the behavior of all everyday objects—the *gravitational* force and the *electrical* force. All of the various forces mentioned above are actually only different manifestations of these two fundamental forces. The reason is that all matter possesses two important characteristics—*mass* and *electrical charge*. The idea that matter possesses mass is familiar, and so the gravitational interaction of all matter is easy to accept. But it is sometimes not fully realized that the basic constituents of matter—*atoms*—are composed of electrically charged units, electrons and nuclei. It is this characteristic of atoms that leads to *all* of the everyday, nongravitational forces that we encounter. When we pull on a rubber band, for example, we must exert a muscular force to overcome the attractive electrical force that exists between the negative charges (the electrons) and the positive charges (the nuclei) in the rubber. The muscular force that we use is also electrical in origin, resulting from the interaction of positive and negative charges in the muscle fibers.

In this chapter we examine in detail the gravitational force and that part of the electrical force that acts between charges at rest—the *electrostatic* force. However, these forces are not sufficient to describe *nuclear* phenomena. The study of processes involving nuclear and elementary particles has shown that there are two additional forces in Nature, the so-called *strong* nuclear force and the *weak* force. Because the gravitational and electrostatic forces have *long ranges* (that is, they are effective over large distances), these forces are exclusively responsible for all large-scale phenomena—those that we encounter in everyday experience and those that we can see taking place in distant stars and galaxies. The nuclear force and the weak force, on the other hand, are of extremely short range and these forces are important only on the scale of nuclear sizes. Nevertheless, these forces play a crucial role in our existence. All life on Earth is sustained by the light that we receive from the Sun and this light is the end result of nuclear processes that take place deep in the Sun's interior. Thus, none of the four basic forces is superfluous; all are required for the orderly operation of the Universe. Of course, there is always the possibility that Nature is more complicated than we are aware of, but at the present time we know of only four basic forces—the gravitational force, the electrical force, the strong nuclear force, and the weak force. There seems to be no need to invoke any additional type of force to account for any observed process. Nature, in her simplicity, has contrived to run her entire Universe by employing only *four* basic forces.

4.1 *The Gravitational Force*

NEWTON'S CALCULATION

The acceleration of an object can only be produced by the application of a force. *Gravitational force* produces an acceleration of 32 ft/sec² for objects falling freely near the surface of the Earth. Although this force is clearly in operation on the Earth, we can ask the question, "Is this force unique to the Earth or does it operate throughout the Universe?" Newton addressed himself to this question but he knew of no way to investigate the "universality" of gravitation except within the solar system. He therefore chose the Earth-moon system for his study and hypothesized that the force that holds the moon in its orbit around the Earth is the same as the force that attracts objects near the surface of the Earth. Newton knew, from previous measurements, that the Earth-moon separation r_m is approximately 60 times the radius of the Earth R_E, which is about 4000 mi (see Fig. 4.1);

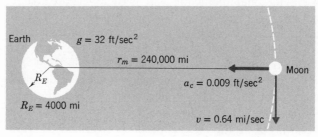

Fig. 4.1 *The centripetal acceleration of the moon in its orbit is due to the gravitational attraction of the Earth.*

therefore, $r_m \cong 240,000$ mi. He also knew the period of the moon's rotation around the Earth — $\tau = 27.3$ days. In order to maintain this motion, the moon must experience a centripetal acceleration equal to $a_c = v^2/r_m$. The linear velocity of the moon is

$$v = \frac{\text{circumference of orbit}}{\text{period of rotation}} = \frac{2\pi r_m}{\tau}$$

$$= \frac{2\pi \times (240,000 \text{ mi})}{(27.3 \text{ days}) \times (86,400 \text{ sec/day})} = 0.64 \text{ mi/sec}$$

Therefore, the centripetal acceleration of the moon is (see Eq. 2.23)

$$a_c = \frac{v^2}{r_m} = \frac{(0.64 \text{ mi/sec})^2}{240,000 \text{ mi}} \times (5280 \text{ ft/mi}) = 0.009 \text{ ft/sec}^2$$

This acceleration is smaller than the value of g at the surface of the Earth by the factor

$$\frac{a_c}{g} = \frac{0.009 \text{ ft/sec}^2}{32 \text{ ft/sec}^2} \cong \frac{1}{3600}$$

The rule that governs the diminution of a quantity as it spreads out uniformly from the source without being absorbed (for example, *light*) is that the intensity decreases with the *square* of the distance:

$$\text{Intensity} \propto \frac{1}{r^2}$$

Newton therefore made the assumption that gravity, just as light, follows the inverse square law of intensity. Thus, if the moon is 60 times farther away from the center of the Earth than is an object at the surface of the Earth, the gravitational force, and hence the acceleration, should be smaller for the moon by a factor $1/(60)^2$ or $1/3600$. This is just the value of the ratio a_c/g.

Although the calculated and observed accelerations agree, thus confirming Newton's hypothesis, it is important to realize two points regarding the derivation:

1. Newton *assumed* that the force of gravity was inversely proportional to the square of the separation of the interacting bodies. This assumption gave the correct result for the Earth-moon system, but there was no guarantee that it would be valid for other systems with different separations. Newton went on to apply the $1/r^2$ law to planetary motion and again found it valid. In more recent times, binary star systems (that is, two stars rotating around one another) have been found to obey the Newtonian form of the gravitation law. It therefore appears that the gravitational force varies as $1/r^2$ throughout the Universe. No exceptions to this rule have ever been found. (However, see the remarks at the end of Section 4.2.)

2. In comparing the centripetal acceleration of the moon with the value of g, both distances were measured from the *center of the Earth*. When he originally made his calculation (in 1666), Newton could not justify this choice and therefore refrained from publishing his results. In fact, it was not until 1687, by which time Newton had developed the necessary mathematical tools (the *calculus*) to justify his earlier calculation, that his law of gravitation was formally announced. Newton had succsseeded in showing, with his calculus, that if two uniform spherical objects attract each other with a force that varies as $1/r^2$, the force will always be correctly calculated if the mass of each body is considered to be *concentrated entirely at its center*. This important result was the key to the universal gravitation law—it showed that all gravitational calculations could be made by considering the entire mass of a spherical body, such as the Earth or the moon, to be located at its center.

Newton's results are summarized by the statement that *the gravitational force which two particles or spherical objects mutually exert on one another is inversely proportional to the square of the distance between their centers and directly proportional to the product of their masses.* That is,

$$F_G = G\frac{m_1 m_2}{r^2}$$
(4.1)

where the constant of proportionality G is called the *universal gravitational constant.* The value of G is (see Section 4.3)

$$G = 6.673 \times 10^{-11} \text{ N-m}^2/\text{kg}^2$$
(4.2)

Example **4.1**

With what force does the Earth attract the moon?

From Tables 1.5 and 1.9 (or from the table of astronomical data inside the front cover) we have the following values:

$r_m = 3.84 \times 10^8$ m
$m_m = 7.35 \times 10^{22}$ kg
$m_E = 5.98 \times 10^{24}$ kg

Therefore, using Eq. 4.1, we have

$$F_G = G\frac{m_m m_E}{r_m^{\,2}}$$

$$= (6.67 \times 10^{-11} \text{ N-m}^2/\text{kg}^2) \times \frac{(7.35 \times 10^{22} \text{ kg}) \times (5.98 \times 10^{24} \text{ kg})}{(3.84 \times 10^8 \text{ m})^2}$$

$$= 2.0 \times 10^{20} \text{ N}$$

We can obtain this result in another way by using the value of the moon's centripetal acceleration calculated above. The force on the moon is just its acceleration times its mass:

$$F_G = m_m a_c = (7.35 \times 10^{22} \text{ kg}) \times (0.009 \text{ ft/sec}^2) \times (0.3048 \text{ m/ft})$$
$$\cong 2 \times 10^{20} \text{ N}$$

4.2 *Planetary Motion*

KEPLER'S LAWS

Between 1609 and 1611, Johannes Kepler enunciated his famous three laws of planetary motion. Kepler's conclusions were based on his analysis of the extensive data relating to planetary positions (particularly pertaining to the planet Mars), that had been acquired by Tycho Brahe during many years of observation.

THE BASIC
FORCES OF
NATURE

Mount Wilson and Palomar Observatories

Fig. 4.2 *A globular cluster of stars in the constellation Hercules. Hundreds of thousands of stars are held together in this cluster by gravitation. Richard Feynman, winner of a share of the 1965 Nobel Prize in physics, has said that "if one cannot see gravitation acting here, he has no soul."*

The statements of Kepler's laws are:

I. The motion of a planet is an ellipse with the Sun at one focus.

II. The line connecting the planet with the Sun sweeps out equal areas in equal times.

III. The period of a planet's motion and its distance from the Sun are related by $R^3/\tau^2 =$ constant, where the constant is the same for all planets.

Only the first of these laws will be of concern in this book.

By geometrical construction from his position data, Kepler showed that planetary orbits are elliptical (but nearly circular). A simple method for constructing an ellipse is the following. First, select the two points, F_1 and F_2, that are the *foci* of the ellipse (Fig. 4.3). Next, attach the ends of a length

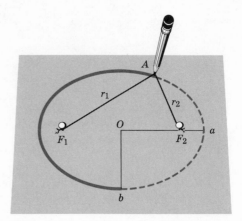

Fig. 4.3 *An ellipse is formed by the line connecting all points (such as A) for which* $r_1 + r_2 =$ *constant.* $\overline{Oa}$ *is the semi-major axis and* $\overline{Ob}$ *is the semi-minor axis.*

of string (which is longer than the distance from F_1 to F_2) to pins at F_1 and F_2. Then, with the tip of a sharp pencil, extent the string until it is taut. The pencil tip will be at a distance r_1 from F_1 and a distance r_2 from F_2 (Fig. 4.3). Finally, the ellipse is constructed by drawing the curve which

Table **4.1** *Planetary Data*

Planet	Semi-major axis (A.U.*)	Period (years)	Mass (Earth masses)	Diameter (Earth diameters)
Mercury	0.387	0.241	0.054	0.37
Venus	0.723	0.615	0.814	0.96
Earth	1.000	1.000	1.000	1.00
Mars	1.524	1.880	0.107	0.52
Jupiter	5.204	11.865	317.4	10.95
Saturn	9.580	29.650	95.0	9.13
Uranus	19.141	83.744	14.5	3.73
Neptune	30.198	165.95	17.6	3.52
Pluto	39.439	247.69	0.18(?)	0.47

*1 astronomical unit (A.U.) is defined as the average distance from the Earth to the Sun: 1 A.U. $= 1.50 \times 10^{11}$ m.

THE BASIC FORCES OF NATURE

the pencil follows as it is moved in such a way that maintains the tautness of the string. The corresponding definition is: an ellipse is the curve that is formed by connecting all points of equal values of the sum $r_1 + r_2$, where r_1 and r_2 are the distances from a given point on the curve to the foci. According to Kepler, all planets move in ellipses with the Sun located at one of the focal points. (The other focal point has no physical significance for the case of planetary motion.)

The ellipse is the general form of the orbit for *bound motion*. (*Bound motion* is the case in which two objects are bound together by their mutual gravitational attraction and cannot escape from each other.) The general form of the orbit for *unbound* motion, as shown by Newton, is the *hyperbola*. For example, if two isolated stars move toward each other, as in Fig. 4.4, they will execute hyperbolic orbits relative to one another.

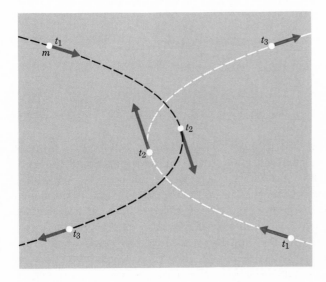

Fig. 4.4 *The hyperbolic orbits of two "colliding" stars that interact via a gravitational force. The positions and the velocity vectors of the two stars are shown for the times t_1, t_2, and t_3.*

DOES THE GRAVITATIONAL FORCE DEPEND EXACTLY ON $1/r^2$?

We have been using Newton's universal gravitational force law with a dependence on distance of the form $1/r^2$. But how do we know that the exponent is not 2.000 001 or 1.999 999? Is the exponent *exactly* 2? A sensitive test of a possible deviation of the exponent from 2 can actually be made by observations of planetary orbits.

Newton showed that if the gravitational force varies exactly as $1/r^2$, then the elliptical orbits described by planets must remain in *fixed* positions. In particular, the point of the ellipse that is closest to the Sun (called the *perihelion*) must remain fixed in its relation to the distant "fixed" stars. Of course, there are small deviations from exact elliptical orbits (perturbations) due to the influence of the other planets, but these deviations are small because of the dominant gravitational force of the Sun; furthermore, mathematical methods exist for the precise calculation of these perturbations. Therefore, if any motion of the perihelion (apart from that expected due to other planets) is observed, this would indicate that the exponent in the force law expression is not *exactly* 2.

About a hundred years ago, a small unexplained motion in the perihelion of Mercury was observed. The perihelion moves forward (*precesses*) at a very slow rate so that the orbit has the appearance of a slowly rotating ellipse (Fig. 4.5). After subtraction of all of the effects due to other planets, there is a net precession that amounts to 43 sec of arc per century. That is, it would require more than 400 yr (or ~ 1600 revolutions) for the perihelion to move

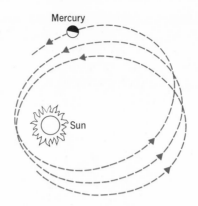

Fig. 4.5 *The perihelion of Mercury's orbit is observed to precess slowly. (The elongation of the orbit and the amount of precession have been greatly exaggerated.)*

by an amount equal to the diameter of the moon (0.5 degree = 1800 arc sec).

At first, the precession of Mercury's perihelion was thought to indicate the presence of an unobserved planet near the Sun that would influence the motion of Mercury, and whose effects would not have been included in the previous perturbation calculations. This planet (prematurely named *Vulcan*) was sought for many years without success. Early in this century, Einstein showed that as a consequence of his relativity theory, there is a small correction needed in the Newtonian form of the gravitational force law. This correction depends on the planet's velocity, and therefore, is important only for Mercury, which has the highest velocity of any planet. With this correction, the precession of Mercury's perihelion is entirely accounted for.

Although a small relativistic correction in Newton's form of the gravitational force law is required for objects in motion with high velocities, the Newtonian equation is valid for essentially all practical purposes and it appears to be entirely correct for *static* gravitational forces.

4.3 *The Electrostatic Force*

POSITIVE AND NEGATIVE ELECTRICITY

It has been known for about 200 years that in Nature there are two basic types of electricity which we designate as *positive* and *negative*. The basic carriers of negative electricity are, of course, *electrons,* which constitute the outer portions of all atoms. The atomic cores, the nuclei, are the seats of positive electricity, the carriers of which are *protons*. All macroscopic matter is basically electrically *neutral*, because the magnitude of the negative electrical charge carried by an electron is equal to that of the positive electrical charge carried by a proton and all atoms in their natural states contain equal numbers of protons and electrons.

The distribution of electrical charge is almost always accomplished by the movement of *electrons;* the more massive positively-charged nuclei remain essentially stationary in almost all electrical processes. That is, a material is given a negative charge by the addition of excess electrons or

THE BASIC
FORCES OF
NATURE

84

a positive charge by the removal of electrons; the number of *atoms* is not changed in either case.

A basic property of electricity is that *like charges repel* and *opposite charges attract*. These facts can easily be demonstrated by some simple experiments. When a glass rod is rubbed with a silk cloth, the friction between the materials causes electrons to be transferred from the glass to the silk. Thus, the glass rod becomes *positively* charged. When a rubber rod is rubbed with a piece of fur, electrons are transferred to the rubber rod and it becomes *negatively* charged. If we suspend a pair of light-weight balls (such as pith balls) on strings, as in Fig. 4.6, and touch each of the balls with a glass rod that has been rubbed with a silk cloth, the balls will both acquire positive charges and will repel each other. On the other hand, if one of the balls has been touched instead with a rubber rod that has been rubbed with fur, it would acquire a negative charge and would therefore be attracted to the other ball.

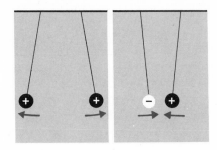

Fig. 4.6 *Like charges repel; opposite charges attract.*

These experiments clearly establish the rules for attraction and repulsion. They also demonstrate that the glass rod carries a charge opposite to that carried by the rubber rod, but other experiments are necessary to show which is positively charged and which is negatively charged (according to the convention that electrons carry negative charge).

Certain types of materials (such as metals) have an interesting property—a fraction of the atomic electrons are not bound to any particular atom but are free to move about in the material. Such materials are termed *conductors*. If an electrical charge is placed on such a material the mutual repulsion will rapidly cause the charge to distribute itself over the surface. On the other hand, if an electrical charge is placed on an *insulating* material, the charge will remain localized. Insulators (for example, glass, plastics) do not have free electrons and therefore electricity does not readily flow in such materials.

THE CONSERVATION OF CHARGE

An object can be given an electrical charge by the transferral of electrons to or away from the object. Charge may be *lost* by the object but it is then *gained* by some other object. It is one of the fundamental conservation laws of Nature that *the total amount of electrical charge in an isolated system remains constant*. No deviation from the law of charge conservation has ever been discovered.

COULOMB'S LAW OF ELECTROSTATIC FORCE

In 1785, the French physicist Charles Augustin de Coulomb (1736–1806) extensively studied electrostatic forces. From his measurements, Coulomb concluded that the electrostatic force between two charged objects varies as the inverse square of the distance between them.[1] That is, the electrostatic force has the same dependence on distance as does the gravitational force. The electrostatic force is proportional to the product of the charges involved (again, of the same general form as the gravitational case with, of course, *charge* substituted for *mass*). Therefore, the expression for the electrostatic force law (also called *Coulomb's law*) is

$$F_E = -k\,\frac{q_1 q_2}{r^2} \qquad (4.3)$$

where q_1 and q_2 are the charges on the two objects and r is their separation. Notice that the expression for F_E contains a *negative* sign, whereas Eq. 4.1 for F_G does not. The reason for the introduction of the sign difference is to maintain the convention that a *positive* force is *attractive*, whereas a *negative* force is *repulsive*. The expressions for F_E and F_G must therefore differ in sign because mass is always positive and the gravitational force is always attractive. Like charges repel and therefore, when q_1 and q_2 have the same sign, F_E must be negative to indicate a repulsive force. Similarly, when the charges have opposite signs, their product is negative which cancels the overall negative sign and produces a positive (attractive) force.

Notice also in Eq. 4.3 that we have an "electrostatic force constant" analogous to G for the gravitational case. The dimensions of force and distance have already been specified, but those of charge and the constant k have not. Therefore, we have some flexibility in choosing the units for q and for k. The procedure is to specify the size of the unit for charge in terms of the charge on the basic carrier of negative electric charge, the *electron*. We call the unit of charge the *coulomb* (C) and in terms of this unit the magnitude of the electron charge is

$$e = 1.602 \times 10^{-19}\,\text{C} \qquad (4.4)$$

(The electron charge is *negative,* and the symbol e represents only the *size* of this basic charge unit.)

The coulomb is a very large unit since it corresponds to an excess (or defecit) of approximately 6×10^{18} electrons.

With the unit of charge defined in this way, the value of k is then determined. The value is

$$k = 9.0 \times 10^9\,\text{N-m}^2/\text{C}^2 \qquad (4.5)$$

That is, to calculate the electrostatic force in newtons between two charges by using Eq. 4.3, the charges must be given in coulombs, the separation must be given in meters, and k must be set numerically equal to 9.0×10^9.

---

[1] Coulomb concluded that the power of the distance was 2 ± 0.02. Measurements made recently at Princeton University have shown that the exponent is 2 to within 1 part in 10^{12}.

Example **4.2**

Calculate the electrostatic force *on* q_1 due to q_2 (F_{12}) and the force *on* q_2 *due to* q_1 (F_{21}) for the case illustrated in the figure.

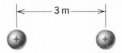

$$q_1 = +3 \times 10^{-3}\,C \qquad q_2 = +5 \times 10^{-3}\,C$$

Using Coulomb's law (Eq. 4.3),

$$F_{12} = -(9.0 \times 10^9) \times \frac{(+3 \times 10^{-3}) \times (+5 \times 10^{-3})}{(3)^2}$$

$$= -1.5 \times 10^4\,N \text{ (i.e., to the } \textit{left}\text{)}$$

$$F_{21} = -(9.0 \times 10^9) \times \frac{(+5 \times 10^{-3}) \times (+3 \times 10^{-3})}{(3)^2}$$

$$= -1.5 \times 10^4\,N \text{ (i.e., to the } \textit{right}\text{)}$$

Both forces have the same magnitude (Newton's third law applied to electrostatic forces) and the negative sign for each indicates that the force on each charge is *repulsive*.

Example **4.3**

Compare the electrostatic and gravitational forces that exist between an electron and a proton.

The electrostatic force law and the gravitational force law both depend on $1/r^2$:

$$F_E = -k\,\frac{q_1 q_2}{r^2}; \quad F_G = G\,\frac{m_1 m_2}{r^2}$$

Therefore, the ratio F_E/F_G is independent of the distance of separation:

$$\frac{F_E}{F_G} = -\frac{kq_1 q_2}{Gm_1 m_2}$$

For the case of an electron and a proton this becomes

$$\frac{F_E}{F_G} = \frac{k(-e)(+e)}{Gm_e m_p} = \frac{ke^2}{Gm_e m_p}$$

Substituting the values of the quantities, we find

$$\frac{F_E}{F_G} = \frac{(9.0 \times 10^9\,\text{N-m}^2/\text{C}^2) \times (1.6 \times 10^{-19}\,\text{C})^2}{(6.67 \times 10^{-11}\,\text{N-m}^2/\text{kg}^2) \times (9.11 \times 10^{-31}\,\text{kg}) \times (1.67 \times 10^{-27}\,\text{kg})}$$

$$= 2.3 \times 10^{39}$$

Thus, the electrostatic force between elementary particles is enormously greater than the gravitational force. Therefore, only the electrostatic force

4.3

THE
ELECTROSTATIC
FORCE

is of importance in atomic systems. At very small distances the strong nuclear force overpowers even the electrostatic force and is the dominant force in all nuclear systems.

4.4 The Nuclear and Weak Forces

WHAT HOLDS NUCLEI TOGETHER?

The two forces that we have described thus far—the gravitational and electrical forces—are the only forces needed to account for the motions of all everyday objects and even the behavior of atomic systems. However, when we look deeper into the atom and probe into the nature of the forces that operate within nuclei, we find that gravitational and electrical forces are no longer adequate to describe the effects that are observed.

We know that nuclei are extremely small—typical radii are only a few times 10^{-15} m. These nuclei contain positive charges (up to $\sim 100\, e$) and we know that electrostatic forces, especially at small distances, can be quite large. Obviously, there must be some extremely strong attractive force that acts within nuclei and overcomes the repulsive Coulomb force that tends to repel the protons from one another.

Example **4.4**

Calculate approximately the electrostatic force that exists between two protons in a typical nucleus.

The radius of an iron nucleus is about 5.4×10^{-15} m. Let us take 2×10^{-15} m as a typical separation for two protons in this nucleus. Then, the electrostatic force between the proton is

$$F_E = -k\frac{e^2}{r^2} = \frac{-(9.0 \times 10^9 \text{ N-m}^2/\text{C}^2) \times (1.6 \times 10^{-19} \text{ C})^2}{(2 \times 10^{-15} \text{ m})^2} \cong 60 \text{ N}$$

This is an enormous repulsive force; it is approximately equal to the gravitational force on a 6-kg mass near the surface of the Earth! This result emphasizes the extraordinary strength of nuclear forces since these forces must overcome the electrostatic repulsion and hold the nucleus together.

THE STRONG NUCLEAR FORCE

The *strong nuclear force* acts at small distances within nuclei and maintains the stability of nuclei in spite of their tendency to fly apart because of Coulomb repulsion. (The designation "strong" is to distinguish this force from the *weak* force that acts between nuclear and elementary particles, which we shall describe below.)

The strong nuclear force acts attractively between protons and protons (*p-p*), between protons and neutrons (*p-n*), and between neutrons and neutrons (*n-n*). We know that the *p-n* and *n-n* forces are essentially identical, and, apart from the Coulomb portion, the *p-p* force is the same as the *p-n* and *n-n* forces. Since protons and neutrons have so many similar properties (except primarily for the lack of charge on the neutron), these particles are

THE BASIC
FORCES OF
NATURE

88

collectively called *nucleons*. Therefore, when we discuss the *nucleon-nucleon* (*N-N*) force, we include the three possible combinations listed above.

The *N-N* force has a dependence on distance that is distinctly different from that of the gravitational and electrostatic forces. These latter forces vary as $1/r^2$ and therefore are termed *long-range* forces; that is, gravitational and electrical effects are manifest over large distances. The *N-N* force, on the other hand, is *short-range* and is effective only over distances up to $\sim 10^{-15}$ m (nuclear dimensions). When a proton is removed from a nucleus, by the time a separation distance of a few times 10^{-15} m has been reached, there is no longer any nuclear force attracting the proton toward the nucleus—only the repulsion of the Coulomb force remains.

The *N-N* force is not entirely attractive. At a distance of $\sim 0.5 \times 10^{-15}$ m, the attraction begins to decrease, and for distances smaller than $\sim 10^{-16}$ m, the *N-N* force becomes *repulsive*. That is, the *N-N* force has a *hard core* that resists pushing two nucleons too close together. The dependence of the *N-N* force on distance is shown schematically in Fig. 4.7. It is possible to show this force only *schematically* because it is not a simple central force[2] as are the gravitational and electrostatic forces. There are important quantum mechanical effects that are crucial in the *N-N* case and make the detailed description of the *N-N* force extremely complicated. We do not yet know, in fact, all of the details of the *N-N* force; this is one of the central problems facing nuclear physics today.

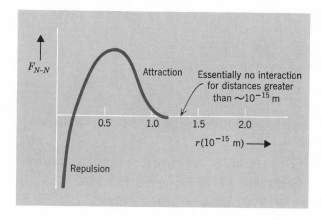

Fig. 4.7 *The nucleon-nucleon force is strongly attractive, but only in the range from $\sim 10^{-16}$ to $\sim 10^{-15}$ m. That is, F_{N-N} is positive (attractive) for this range of r and is negative (repulsive) for r less than $\sim 10^{-16}$ m.*

THE WEAK FORCE

The process of nuclear β decay involves (Section 12.2) the emission from the parent nucleus of an electron and a massless particle called a *neutrino*. Just as a proton and a neutron interact primarily by the *strong* nuclear force, the electron (*e*) and the neutrino (*v*) interact (*exclusively*) by the *weak* force. This weak force is responsible for nuclear β decay and certain other processes involving elementary particles. Our understanding of the weak force is at present very meager; we have no detailed knowledge, for example, of the

---

[2] A *central force* between two particles is one that acts only along the line that connects the particles. The *N-N* force has parts that are central and parts that are non-central.

dependence of the force on distance, except that it is of short range, certainly no greater than that of the strong nuclear force.

A weak force exists between *all* pairs of elementary particles. It is the exclusive force between electrons and neutrinos, but the same force (albeit much weaker than the electrical or strong nuclear force) exists even between two protons. The relative sizes of the various forces that act between pairs of elementary particles are listed (in order of magnitude) in Table 4.2. The nucleon-nucleon force is arbitrarily assigned unit magnitude.

Table **4.2** *Comparison of the Forces that Act between Elementary Particles*

Force	Ratio of Strengths at Small Distances ($\sim 10^{-15}$ m)			
	e-v	*e-p*	*p-p*	*p-n* *n-n*
Strong nuclear	0	0	1	1
Electrostatic	0	10^{-2}	10^{-2}	0
Weak	10^{-13}	10^{-13}	10^{-13}	10^{-13}
Gravitational	0	10^{-41}	10^{-38}	10^{-38}

Because the neutrino interacts with all other elementary particles only through the weak force and because neutrinos can now be produced in copious amounts at large accelerator installations, the study of neutrinos and the weak interaction is at present being extensively pursued.

Summary of Important Ideas

There are only four basic forces in Nature—gravitational, electrical, strong nuclear, and weak nuclear.

The gravitational and electrical forces depend on $1/r^2$ and are *long-range* forces. The nuclear force and the weak force have *short ranges*.

Planets move in *elliptical* orbits around the Sun.

For purposes of gravitational calculations, the entire mass of a uniform spherical object can be considered to be concentrated at its *center*. For purposes of electrical calculations, the entire charge of a uniform spherical distribution of charge can be considered to be concentrated at its *center*.

The net electrical charge of an isolated system remains *constant*.

The *strong* nuclear force is responsible for the stability of nuclei. The *weak* force is responsible for nuclear β decay.

The order of *strengths* of the basic forces is (1) strong nuclear, (2) electrical, (3) weak, and (4) gravitational.

THE BASIC
FORCES OF
NATURE

Questions

4.1

The mass of the moon is approximately 1/81 of the mass of the Earth but the gravitational attraction on an object at its surface is only about 1/6 of the gravitational attraction on the same object at the surface of the Earth. Why?

4.2

Comets are members of the solar system. Explain why some comets (such as Halley's comet) appear remarkably bright for a brief period and then become unobservable, even with powerful telescopes, for a number of years.

4.3

Two protons approach each other in space. Describe their relative orbits if they approach (a) head-on and (b) with initially parallel but displaced paths.

4.4

A proton is moving directly toward a certain atom. Describe the forces that the proton experiences as a function of its distance from the nucleus of the atom.

4.5

What forces act between the following pairs of elementary particles: v-p, v-n, e-n?

Problems

4.1

What is the gravitational attractive force between an object with a mass of 40 kg and an object with a mass of 90 kg if the centers of the two objects are separated by the distance of 6 m?

4.2

Near the surface of the Earth, we know that the acceleration due to gravity is $g = 32$ ft/sec^2. What values would be found at distances above the Earth's surface of 4000 mi and 8000 mi? (The radius of the Earth is 4000 mi.)

4.3

With what force does the Sun attract the Earth?

4.4

The gravitational force on an object of mass m at the surface of the Earth is $F = GmM/R^2$, where M is the mass and R is the radius of the Earth. This force is also equal to the *weight* of the object, $F = W = mg$. From these two expressions for F calculate the mass of the Earth, given the values for G, R, and g.

4.5

What is the value of g at the surface of the Sun? (Refer to the equation in Problem 4.4.)

4.6

Two identical spherical objects are separated by a distance of 1 m. If the gravitational force between them is 6.7×10^{-9} N, what is the mass of each object?

4.7

A certain object carries a charge of $+4 \times 10^{-5}$ C. Another object, at a distance of 3 m, carries a charge Q and is attracted toward the first object with a force of 8 N. What is the charge Q?

4.8

Two identical spherical copper balls carry charges of $+2.8 \times 10^{-4}$ C and -4×10^{-5} C and their centers are separated by 2 m. If the balls are brought together until they *touch* and then returned to their original positions, what will be the electrostatic force between them?

4.9

A proton and an electron are at rest a distance 10^{-10} m apart. What is the electrostatic force exerted by one of the particles on the other? When they begin to move under the influence of this force, what is the initial acceleration of each particle?

CHAPTER 5
ENERGY

Without a doubt, *energy* is the single most important physical concept in all of science. Historically, there grew up intuitive ideas regarding *energy* and the closely associated concept of *work* just as for length, time, and mass. Some of the ideas of the layman are, in fact, quite closely related to the precisely formulated ideas of the physicist. It could be said, for example, "Eat a good meal and you will have a lot of energy," and "A person who has a great deal of energy can do a large amount of work." These statements correspond closely to those that the physicist would make: "The stored chemical energy in foodstuffs can be transferred to biological systems," and "Energy is the capacity to do work."

The importance of the concept of energy lies in the fact that various forms of energy within an isolated system can be transformed into one another *without a change in the total amount of energy*. That is, in any physical process, *energy is conserved*. The law of energy conservation, which is obeyed in every known process, is the most fruitful law in physics for the analysis of phenomena of every sort.

5.1   *Work*

MOTION AGAINST A FORCE

If we exert a force on an object and move it a certain distance, we say that we have done *work*. Lifting an object, for example, requires exerting a force sufficient to overcome the downward gravitational force. If we apply such a force sufficiently long to raise the object to a height *h*, we have done a certain amount of work. Or, if we push an object across a rough surface with a force sufficiently large to overcome friction and move it a distance *s*, we have again done a certain amount of work. The amount of work done is proportional to both the applied force and the distance through which the force acts.

Suppose that we apply a constant force to an object and thereby cause it to move with *constant velocity* over a rough surface. Since the acceleration of the object is zero, we know that there is no *net* force being applied. The externally applied force in this case just balances the retarding frictional force so that $F_{\text{net}} = 0$. Nevertheless, work *is* being done by the external force because the external force is working against the frictional force. The act of displacing an object against a retarding force (such as friction or gravity) by the application of an external force constitutes *work*.

WORK = FORCE × DISTANCE

Work is the product of the force applied to an object by an outside agent and the distance through which the force acts on the object. In Fig. 5.1, the work done by the force F in displacing the block a distance s is

$$W = Fs. \tag{5.1}$$

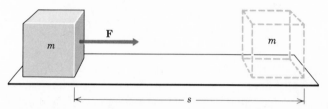

Fig. 5.1 *When the displacement is in the same direction as the applied force, the work done is $W = Fs$.*

In this case the direction of the displacement is the same as the direction of the applied force. In Fig. 5.2, however, the force is applied at an angle relative to the direction of motion. This force can be considered to be the vector sum of two independent forces, F_x and F_y. The component F_x has the direction of the displacement; therefore the work done by this compo-

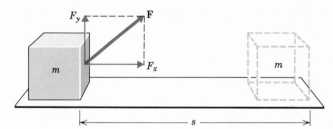

Fig. 5.2 *Only the component of the force in the direction of the displacement does any work, $W = F_x s$.*

nent of the force is $W = F_x s$. There is no displacement in the y-direction; therefore, the component F_y *does no work*. (As long as F_y is less than the downward gravitational force, no motion in the vertical direction can result). In general, then, our definition of work must be modified:

Work is the product of the component of force in the direction of the displacement and the magnitude of the displacement produced by the force.

Although work depends on the product of two quantities that are vectors, work is a *scalar* quantity.

ENERGY

In the metric system of units, the dimensions of *work* are the dimensions of *force* times the dimensions of *distance,* or N-m. To this unit we give the special name *joule* (J):

$$1 \text{ N-m} = 1 \text{ kg-m}^2/\text{sec}^2$$
$$= 1 \text{ joule (J)}$$

Example **5.1**

A horizontal force of 5 N is required to maintain a velocity of 2 m/sec for a box of mass 10 kg sliding over a certain rough surface. How much work is done by the force in 1 min?

First, we must calculate the distance traveled:

$$s = vt = (2 \text{ m/sec}) \times (60 \text{ sec}) = 120 \text{ m}$$

Then,

$$W = Fs = (5 \text{ N}) \times (120 \text{ m}) = 600 \text{ N-m} = 600 \text{ J}$$

5.2   *Kinetic Energy*

THE ENERGY ASSOCIATED WITH MOTION

Suppose we have a *free* object—for example, an object in space initially at rest (with respect to some coordinate system) and subject to no forces. If we apply to the object a constant force F, the object will be accelerated. After moving a distance $s = \frac{1}{2}at^2$, the object will have a velocity $v = at$. The work done on the object is

$$W = F \times s = (ma) \times (\tfrac{1}{2}at^2) = \tfrac{1}{2}m \times (at)^2$$
$$= \tfrac{1}{2}mv^2 \qquad (5.2)$$

Thus, an amount of work W has been performed on the object and we say that the object has acquired an amount of *energy* equal to $\frac{1}{2}mv^2$. This energy, which the object possesses *by virtue of its motion,* is called *kinetic energy.*

$$\boxed{\text{Kinetic energy: } KE = \tfrac{1}{2}mv^2} \qquad (5.3)$$

Clearly, the unit of kinetic energy is the same as that for work.

Example **5.2**

A free particle, which has a mass of 2 kg, is initially at rest. If a force of 10 N is applied for a period of 10 sec, what kinetic energy is acquired by the particle?

In order to calculate the kinetic energy we must first compute the final velocity acquired by the particle. Using $a = F/m$, we have

$$v = at = \left(\frac{F}{m}\right)t = \left(\frac{10 \text{ N}}{2 \text{ kg}}\right) \times (10 \text{ sec}) = 50 \text{ m/sec}$$

Then,

$$KE = \tfrac{1}{2}mv^2 = \tfrac{1}{2} \times (2 \text{ kg}) \times (50 \text{ m/sec})^2 = 2500 \text{ J}$$

How much work was done by the applied force? The distance moved is

$$s = \tfrac{1}{2}at^2 = \tfrac{1}{2}\left(\frac{F}{m}\right)t^2 = \tfrac{1}{2} \times \left(\frac{10 \text{ N}}{2 \text{ kg}}\right) \times (10 \text{ sec})^2 = 250 \text{ m}$$

so that the work done is

$$W = F \times s$$
$$= (10 \text{ N}) \times (250 \text{ m}) = 2500 \text{ J}$$

In this case, the work done is transformed entirely into the kinetic energy of the particle.

5.3 *Potential Energy*

THE ENERGY ASSOCIATED WITH POSITION

Suppose we lift an object of mass m, originally at rest, to a position that is a height h above its initial position and leave the object again at rest. Clearly, we have done work against the gravitational force, but there is no net change in velocity and therefore we have imparted no kinetic energy.

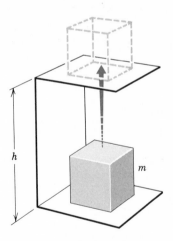

Fig. 5.3 *An object acquires a potential energy mgh by being raised to a height h.*

However, the object *does* possess energy by virtue of its *position*. We can easily see this by allowing the object to fall toward its original position. After falling through a distance h, it will have acquired a velocity $v = \sqrt{2gh}$ (see Example 2.3) and its kinetic energy will be

$$KE = \tfrac{1}{2}mv^2 = \tfrac{1}{2}m(2gh) = mgh$$

This amount of energy, *mgh*, may be accounted for as follows. In order to balance the gravitational force on an object of mass m, the application of a force mg is required. If a force infinitesimally greater[1] than mg acts through a distance h (the height to which the object is raised), an amount of work $W = Fs = mgh$ is done. The object then possesses an energy mgh that has the potential of being transformed into kinetic energy (by falling through the height h). We call this energy, which the body possesses *by virtue of its position*, the *potential energy* of the body. That is, the work done against the gravitational force is stored by the object raised and is retained as potential energy:

$$\boxed{\text{Potential energy: } PE = mgh} \tag{5.4}$$

ENERGY AND WORK

As we have mentioned earlier, *energy is the capacity to do work*. The potential energy acquired by an object in being raised to a certain height can be converted into work in a number of ways. One way of accomplishing this is by means of a *pile-driver* as illustrated schematically in Fig. 5.4. By falling through a height h a block of mass m transforms its potential energy mgh into kinetic energy $\frac{1}{2}mv^2$, where $v = \sqrt{2gh}$. The moving block, in being stopped, can do an amount of work $mgh = \frac{1}{2}mv^2$. In Fig. 5.4, this amount of work is expended in driving the stake into the ground and is equal to $\bar{F} \times \Delta s$, where $\bar{F}$ is a very large average force and Δs is the small distance through which the stake moves.

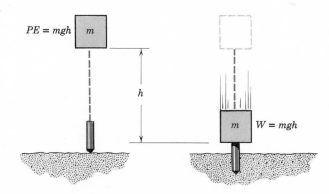

$PE = mgh$

h

m $W = mgh$

Fig. 5.4 *The potential energy of the block is transformed into kinetic energy by falling and does work by driving the stake into the ground.*

Example **5.3**

How much work is required to raise a 0.1-kg block to a height of 2 m and simultaneously give it a velocity of 3 m/sec?

The work done is the sum of the potential energy, $PE = mgh$, and the kinetic energy, $KE = \frac{1}{2}mv^2$:

$PE = mgh = (0.1 \text{ kg}) \times (9.8 \text{ m/sec}^2) \times (2 \text{ m})$
$\quad\quad = 1.96 \text{ J}$

[1]By choosing a force infinitesimally greater than mg, we are assured that we can actually *move* the object (not just balance the gravitational force); but, at the same time, we make no appreciable error by equating the work done to mgh.

$$KE = \tfrac{1}{2}mv^2 = \tfrac{1}{2} \times (0.1 \text{ kg}) \times (3 \text{ m/sec})^2$$
$$= 0.45 \text{ J}$$

$$W = PE + KE$$
$$= 1.96 \text{ J} + 0.45 \text{ J} = 2.41 \text{ J}$$

5.4 Energy is Only Relative

ONLY ENERGY DIFFERENCES ARE IMPORTANT

Unlike length and mass, *energy* has no absolute value. Suppose we ask the question "How much potential energy does a certain body have?" The answer is not simply "*mgh*," because then one can ask "*h* above what?" One could always drop the object into a hole and release some extra potential energy as kinetic energy. Therefore, the potential energy of the object at the surface of the Earth surely is not *zero*. The absolute value of the potential energy at any particular point has, in fact, no physical significance; it is only the *difference* in potential energy between two points that is important. In moving an object between two points, only the difference in potential energy can be converted into kinetic energy. Since we can always add a constant amount to the value of the potential energy at each of the two positions without altering the *difference* in potential energy between the positions, the absolute value of the potential energy is arbitrary. Therefore, in any particular situation we are free to choose for potential energy a zero level that is convenient for the case at hand.

Kinetic energy is also a relative concept. The kinetic energy of a moving automobile, for example, appears to have different values for an observer standing on the road and for an observer in a train traveling on a track alongside the road. It is the *relative* velocity that determines the kinetic energy through the relation $\tfrac{1}{2}mv^2$. The kinetic energy of an object has different values with respect to different moving coordinate systems. Again, it is only the *change* in kinetic energy that is important because it is just this change in energy that appears as work in any frame of reference.

5.5 Conservation of Energy

OUR MOST USEFUL PHYSICAL PRINCIPLE

In the simple examples given in the preceding sections it was implicit that potential energy could be transformed into kinetic energy, and *vice versa*, without any loss of energy. That is, a mass *m* falling through a height *h* acquires a kinetic energy $\tfrac{1}{2}mv^2$, where $mgh = \tfrac{1}{2}mv^2$, or $(PE)_{\text{initial}} = (KE)_{\text{final}}$. Thus, energy (in one form or another) has been *conserved* during the process. This is, in fact, a general result embodied in the principle (or *law*) of *the conservation of energy*.

Energy conservation is a far-reaching principle, and, just as for linear momentum and angular momentum, there is no known exception to the rule that energy is conserved in every physical process. Indeed, we have the attitude that if we do not find a balance of energy in a certain process,

Table **5.1**   *Range of Energies*

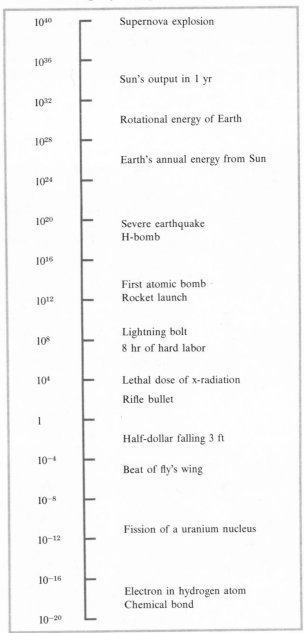

10^{40}	Supernova explosion
10^{36}	
	Sun's output in 1 yr
10^{32}	
	Rotational energy of Earth
10^{28}	
	Earth's annual energy from Sun
10^{24}	
10^{20}	Severe earthquake H-bomb
10^{16}	
	First atomic bomb
10^{12}	Rocket launch
	Lightning bolt
10^{8}	8 hr of hard labor
10^{4}	Lethal dose of x-radiation Rifle bullet
1	
	Half-dollar falling 3 ft
10^{-4}	Beat of fly's wing
10^{-8}	
	Fission of a uranium nucleus
10^{-12}	
10^{-16}	
	Electron in hydrogen atom Chemical bond
10^{-20}	

we invent a form of energy that exactly makes up the deficit! This is not really a trick or a dishonest attempt to cover up our ignorance about Nature, for once we invent a new form of energy we must thereafter use the same definition and always incorporate this new form into our calculations in the same way. If we have made a poor choice, we will rapidly come upon a contradiction. Following this attitude, we have invented, for example, thermal energy, electromagnetic energy, and nuclear energy. Using these ideas

5.5
CONSER-
VATION
OF ENERGY

consistently, we find no contradictions. Energy can take many forms and can be changed from one form to another. But in any physical process, if a strict accounting is made, it is always found that the total amount of energy is *constant*—energy is conserved.

A principle is useful only if it allows us to gain some insight into the way Nature works. From this standpoint, energy conservation is surely the most useful single principle in science. Many problems in which complicated forces are involved and therefore the solutions to which are extremely difficult to construct by using Newton's laws can, nevertheless, be solved in a simple way by using the conservation laws, particularly energy conservation. Nothing, it is said, succeeds like success, and energy conservation is certainly a successful principle.

5.6 *Gravitational Potential Energy*

WORK REQUIRED TO LIFT AN OBJECT

Near the surface of the Earth, where we can consider the gravitational force to be constant, we know that to life an object of mass m through a height h requires an amount of work $W = mgh$. After lifting, we say that the object has a gravitational potential energy $PE_G = mgh$. If we consider distances that are no longer negligible compared to the radius of the Earth, then we must take account of the fact that the gravitational force varies as the square of the distance from the center of the Earth.

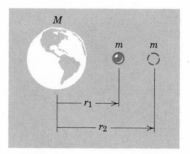

Fig. 5.5 *Work done by an outside agency is required to move m from r_1 to r_2.*

We now calculate the amount of work required to raise an object of mass m from an initial position that is a distance r_1 from the center of the Earth, to a final position that is a distance r_2 from the center of the Earth, without any change in the kinetic energy of the object (see Fig. 5.5). The force on m at r_1 is

$$F_1 = G\frac{Mm}{r_1^2} \qquad \text{at } r_1 \tag{5.5a}$$

and at r_2 the force is

$$F_2 = G\frac{Mm}{r_2^2} \qquad \text{at } r_2 \tag{5.5b}$$

ENERGY

We know that the amount of work done on an object is equal to the average force exerted multiplied by the distance through which the object is moved:

$$W = \bar{F} \times s \tag{5.6}$$

We need a procedure that yields an average force that is closer to F_2 than to F_1 because the force falls off rapidly in going from r_1 to r_2. An *arithmetic* average is therefore not correct in this case; the average appropriate for a $1/r^2$ force is the *geometric* average, namely,

$$\bar{F} = G \frac{Mm}{r_1 r_2} \tag{5.7}$$

(We have not proved that this average is correct; we have only argued that it is reasonable. In spite of this arbitrariness, the result is nevertheless *exact*.) Substituting this expression for $\bar{F}$ and $s = r_2 - r_1$ into Eq. 5.6, we find

$$W_{12} = G \frac{Mm}{r_1 r_2} \times (r_2 - r_1)$$

$$= GMm \left(\frac{1}{r_1} - \frac{1}{r_2} \right) \tag{5.8}$$

The increase in gravitational potential energy of m in moving from r_1 to r_2 is just the work required to effect this change of position; that is,

$$\boxed{PE_G = GMm \left(\frac{1}{r_1} - \frac{1}{r_2} \right)} \tag{5.9}$$

Notice that the gravitational potential energy depends only on the initial and final positions of the mass, r_1 and r_2. Although we considered here a particularly simple path by which the mass was moved from r_1 to r_2, in fact, PE_G depends only on r_1 and r_2 for *any* path that connects the initial and final positions.

ESTABLISHING THE "ZERO" OF POTENTIAL ENERGY

According to our general expression (Eq. 5.9), the increase in gravitational potential energy in raising an object of mass m from the surface of the Earth (radius R_E) to a position that is a distance r from the Earth's center is

$$PE_G = GMm \left(\frac{1}{R_E} - \frac{1}{r} \right) \tag{5.10}$$

The term GMm/R_E is a constant and so does not affect *differences* in PE_G for various values of r. Because the zero level of PE_G is arbitrary, we can subtract this term from the expression for PE_G and write

$$\boxed{PE_G = - \frac{GMm}{r}} \tag{5.11}$$

We see from this equation that $PE_G = 0$ as r becomes indefinitely large, $r \to \infty$. That is, we arbitrarly select the zero level of PE_G to be the situation in which the two masses are separated by an infinite distance. The use of an infinite separation distance as the reference level for PE_G simply means that we measure PE_G relative to a configuration (infinite separation) for which there is no interaction between the objects and, hence, no potential energy.

According to our choice of position for zero gravitational potential energy, the value of PE_G is always *negative* and increases (that is, becomes less negative) as r is increased, becoming zero as $r \rightarrow \infty$. Work is always required to increase the potential energy (that is, to separate the bodies); the work required is a maximum when the bodies are infinitely far from each other in the final position (see Fig. 5.6).

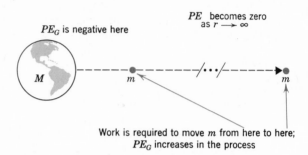

Fig. 5.6 *Work is required to increase the gravitational potential energy from a negative value to a less negative value or to zero (as $r \rightarrow \infty$).*

POTENTIAL ENERGY DIFFERENCES

It must be remembered that only *changes* in potential energy are physically meaningful. Therefore, we are usually concerned with calculating $\Delta(PE_G)$ in moving an object of mass m from r_1 to r_2 as measured from M. According to Eq. 5.8,

$$\Delta(PE_G) = W_{12} = GMm \left(\frac{1}{r_1} - \frac{1}{r_2} \right) \tag{5.12}$$

If $W_{12} > 0$, then work was done by some outside agent *against* the attractive gravitational force. If $W_{12} < 0$, then work was done *by* the gravitational force and this amount of energy can be used, for example, to increase the kinetic energy of the bodies. If two masses with a certain separation are released, the attractive gravitational force will cause the masses to accelerate toward one another and the separation will decrease. Consequently, the gravitational potential energy will decrease, and there will be an equivalent gain in kinetic energy.

TOTAL ENERGY = PE + KE

Figure 5.7a shows a roller-coaster car and the hill-and-valley track over which the car runs. Suppose the car starts from rest at a height h_1 above ground level. From experience we know that the speed of the car will be greatest in the valleys of the track and will be least on the hills. This fact is due to the interchange of potential energy and kinetic energy. Since the potential energy at any point is proportional to the height of that point above the reference (or ground) level, we can convert the track diagram directly into a *potential energy diagram*, as in Fig. 5.7b. From this curve we can read

ENERGY

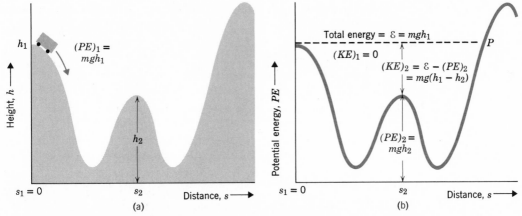

Fig. 5.7 (a) A roller coaster with a car starting from a height h_1. (b) The potential energy diagram corresponding to the roller-coaster of (a). The height of the curve at any position is equal to the PE and the difference between the total energy line and the curve is equal to the KE at that position. Notice that the car cannot pass over the hill on the right since to move above the height of the point P requires an energy greater than the total energy $\mathcal{E}$ of the car.

directly the *PE* at any position. The position $s = s_1 = 0$ corresponds to the starting point where $(PE)_1 = mgh_1$ and $(KE)_1 = 0$. Thus, the total energy $\mathcal{E}$ at $s = s_1$ is $\mathcal{E} = (PE)_1 + (KE)_1 = mgh_1$. If frictional losses are neglected, energy conservation requires that the total energy at any other position is also mgh_1. At $s = s_2$, where the car is at a height h_2, the potential energy is $(PE)_2 = mgh_2$; the kinetic energy at that position must be the difference between $\mathcal{E}$ and $(PE)_2$, or $(KE)_2 = \mathcal{E} - (PE)_2 = mg(h_1 - h_2)$. That is, the kinetic energy at any position is given graphically by the difference between the total energy line and the potential energy curve.

5.7 *Electrostatic Potential Energy*

COMPARISON OF GRAVITATIONAL AND ELECTROSTATIC POTENTIAL ENERGIES

We consider now that portion of electromagnetic energy associated with charges at rest—*electrostatic* energy.

Except for the important fact that electric charge can be either positive or negative, whereas mass is always positive, the case of electrostatic potential energy is the same as that of gravitational potential energy. We have already seen that the gravitational force between two objects with masses m_1 and m_2, separated by a distance r, is

$$F_G = G\frac{m_1 m_2}{r^2}$$

From this fact, we found that the gravitational potential energy in such a case is

$$PE_G = -G\frac{m_1 m_2}{r}$$

where we use the convention that $PE_G = 0$ when $r \to \infty$.

For the electrostatic case, the force between the charges q_1 and q_2, separated by a distance r, is

$$F_E = -k\frac{q_1 q_1}{r^2} \qquad (5.13)$$

where we understand that a *positive* force is *attractive* (that is, when q_1 and q_2 have opposite signs) and a *negative* force is *repulsive* (that is, when the charges have the same sign). Because the expression for F_E is of the same form as that for F_G, it follows that the expression for the electrostatic potential energy PE_E must be of the same form as that for PE_G, except that we must carry along the sign difference between F_G and F_E. Thus,

$$PE_E = -k\frac{q_1 q_2}{r} \qquad (5.14)$$

WORK AND POTENTIAL ENERGY CHANGES

The work done, or change in PE_E, in moving a charge q_1 from r_1 to r_2 as measured from q_2 is

$$W_{12} = \Delta(PE_E) = kq_1 q_2\left(\frac{1}{r_2} - \frac{1}{r_1}\right) \qquad (5.15)$$

Compare this expression with Eq. 5.12 for $\Delta(PE_G)$.

The difference of sign between PE_G and PE_E or between $\Delta(PE_G)$ and $\Delta(PE_E)$ is easy to understand by referring to the concept of *work*. This is illustrated schematically in Fig. 5.8 for the case of electrostatic potential energy. Whenever work is done *against* a gravitational or electrical force, the potential energy *increases*; $W > 0$ and $\Delta(PE) > 0$. If work is done *by* such a force, the potential energy *decreases*; $W < 0$ and $\Delta(PE) < 0$. The first two cases[2] in Fig. 5.8 apply also for the gravitational case.

Example **5.4**

Suppose we have two charges, $q_1 = +4 \times 10^{-4}$ C and $q_2 = -8 \times 10^{-4}$ C, with an initial separation of $r_1 = 3$ m. What is the change in potential energy if we increase the separation to 8 m?

Using Eq. 5.15,

$$\Delta(PE_E) = kq_1 q_2\left(\frac{1}{r_2} - \frac{1}{r_1}\right) = (9 \times 10^9)(+4 \times 10^{-4})(-8 \times 10^{-4}) \times (\tfrac{1}{8} - \tfrac{1}{3})$$
$$= (-2.88 \times 10^3) \times (-\tfrac{5}{24}) = +600 \text{ J}$$

In this case there is a net *increase* in the electrostatic potential energy (that is, $\Delta(PE_E) > 0$) because work was done by an outside agent against the attractive electrostatic force.

---

[2] Only the first two cases since F_G is always *attractive*.

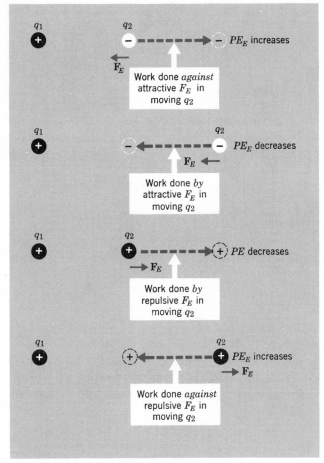

Fig. 5.8 *The changes in PE_E are shown for the four possible cases of a charge (q_2) moved in the presence of a positive charge (q_1).*

5.8 *Potential difference and the Electron Volt*

VOLTS

Equation 5.15 states that the work done in moving a charge q from a distance r_1 to a distance r_2 measured from a fixed charge Q is

$$W = qQ\left(\frac{1}{r_2} - \frac{1}{r_2}\right)$$

If we know q, Q, and the distances, we can always calculate W. If we change q to a new value, then W also changes. However, the ratio W/q does not depend on q; W/q is a *property only of the charge Q*. (We call Q the *source* charge.) The quantity W/q is the work *per unit charge* required to move from r_1 to r_2 in the presence of the electrostatic force due to the source charge

5.8
POTENTIAL
DIFFERENCE
AND THE
ELECTRON
VOLT

Q. This is an extremely useful concept and we therefore give W/q a special name—*potential difference*,[3] V:

Potential difference: $V = \dfrac{W}{q}$ (5.16)

Note that W is the difference in potential energy (for a particular value of q); W/q does not have the dimensions of energy, but of energy (or work) *per unit charge*. If 1 *joule* of work is required to move 1 *coulomb* of charge from r_1 to r_2 in the presence of Q, we say that the potential difference between r_1 and r_2 is 1 *volt* (V). That is,

$1 \text{ V} = 1 \text{ J/C}$ (5.17)

The *volt* is just that unit which is familiar in terms of household electricity. The potential difference between the two wires of common household electrical circuits is 110 V; the potential difference between the terminals of a flashlight battery is 1.5 V.

THE ELECTRON VOLT

A unit of energy that is quite useful in dealing with problems in atomic and nuclear physics is obtained in the following way. Suppose that a charge e, equal to the charge on an electron (which is the same as the charge on a proton, disregarding the sign), is moved from one position to another between which exists a potential difference of 1 V. How much work has been done *on* the charge? Using Eq. 5.16, we have

$$W = qV = e \times V$$
$$= (1.602 \times 10^{-19} \text{ C}) \times (1 \text{ V})$$
$$= 1.602 \times 10^{-19} \text{ J} \qquad (5.18)$$

This unit of *work* or energy is given the special name *electronvolt*, and is denoted by the symbol eV.

$1 \text{ eV} = 1.602 \times 10^{-19} \text{ J}$ (5.19)

Larger units of energy are convenient for many problems, especially in nuclear physics; those most frequently used are

$$1 \text{ kiloelectronvolt (keV)} = 10^3 \text{ eV} = 1.602 \times 10^{-16} \text{ J}$$
$$1 \text{ megaelectronvolt (MeV)} = 10^6 \text{ eV} = 1.602 \times 10^{-13} \text{ J}$$
$$1 \text{ gigaelectronvolt (GeV)} = 10^9 \text{ eV} = 1.602 \times 10^{-10} \text{ J}$$

Example **5.5**

A proton, starting from rest, moves through a potential difference of 10^6 V. What is its final kinetic energy and final velocity?

For the kinetic energy we have, simply,

$$KE = 10^6 \text{ eV} = 1 \text{ MeV} = 1.602 \times 10^{-13} \text{ J}$$

[3]Sometimes, V is called simply the *potential*.

In order to compute the final velocity, we use

$$KE = \tfrac{1}{2}m_p v^2$$

or

$$v = \sqrt{\frac{2KE}{m_p}} = \sqrt{\frac{2 \times (1.60 \times 10^{-13} \text{ J})}{1.67 \times 10^{-21} \text{ kg}}}$$

so that

$$v = 1.38 \times 10^7 \text{ m/sec}$$

If we had considered an electron, instead of a proton, falling through a potential difference of 10^6 V, we could not have computed the final velocity in such a simple way. Relativity theory provides the correct method of calculation. The relativistic effect, which is manifest for a 1-MeV electron, is negligible for a proton of the same energy owing to the much larger mass of the proton. However, for protons with energies of about 100 MeV or more we must also use the relativistic expression for computing the velocity.

The *electronvolt* is commonly used as a unit of energy even in the event that the particle has not fallen through any potential difference. For example, a neutron (which has approximately the same mass as a proton but no electric charge and, therefore, cannot experience an electrostatic force) which is moving with a velocity of 1.38×10^7 m/sec is said to have an energy of 1 MeV. That is, the unit "1 MeV" means that the particle has a definite amount of kinetic energy and it does not matter how the particle acquired this energy.

The eV notation is rarely used for objects larger than atomic size because the unit is too small to be convenient. The energy of a 0.1-g meteorite which strikes the Earth with a velocity of 50 km/sec, for example, has an energy of approximately 8×10^{14} GeV.

5.9   *Heat as a Form of Energy*

THE FIRST LAW OF THERMODYNAMICS

All material things consist of atoms and molecules, and these atoms and molecules are continually in motion, whether the object is gas, liquid, or solid. Therefore, even if an object is motionless in a position of zero potential energy (relative to some base position), energy is associated with the internal motion of the constituent atoms and molecules. That is, there is always a certain *internal energy* for any collection of atoms or molecules. If we alter the system by causing the atoms to move more violently, we say that we have added *thermal energy* to the system (by doing work *on* the system or by adding *heat* to the system), thereby increasing the internal energy. *Thermodynamics* is the branch of physics concerned with the interplay of heat, work, and energy.

In applying the principle of energy conservation in idealized situations we have not previously been concerned with the internal energy of an object. Clearly, our analyses have not been complete, for in any real mechanical system energy is always absorbed by friction. This energy appears as an increase in the internal energy of the objects in the system (thereby increasing the *temperature*), and the *total* energy in the system is conserved.

Let us denote by U the internal energy of a body and by Q the amount of thermal energy (or heat) supplied to the body by friction or some other means. If a body takes in an amount of heat Q, the internal energy is increased by exactly that amount: $\Delta U = Q$. Alternatively, the body could do a certain amount of work when supplied with the heat Q; for example, if the body is a gas, it could expand against a restraining force as in Fig. 5.9. The increase in internal energy is diminished if the body does work

Fig. 5.9 *An amount of heat Q is supplied to a gas and the expanding gas does an amount of work W. The increase in the internal energy of the gas is* $\Delta U = Q - W$. *Notice that a portion of the thermal energy is transformed into gravitational potential energy, i.e., the work done by the expanding gas is* $W = mgh$.

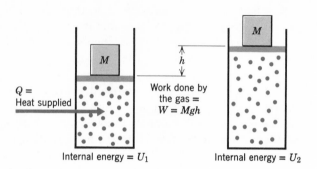

$Q =$ Heat supplied

Work done by the gas = $W = Mgh$

Internal energy = U_1

Internal energy = U_2

when supplied with heat. The principle of energy conservation states that the change in internal energy is equal to the heat supplied *to* the body *minus* the work done *by* the body:

$$\Delta U = Q - W \qquad (5.20)$$

This equation is called the *first law of thermodynamics*, but it is simply a statement of energy conservation when thermal energy is included.

TEMPERATURE

The quantitative specification of internal energy or thermal energy requires the concept of *temperature*. Temperature is a familiar idea, indicating the degree of "hotness" or "coldness" of an object. The difference in the "hotness" of boiling water and the "coldness" of ice is divided into 100 equal parts, called *1 Centigrade degree* (°C). The temperature of melting ice is arbitrarily designated 0°C and that of boiling water 100°C when the measurements are carried out at sea level.

When the temperature of a substance is lowered, the motion of the molecules in the material decreases. In fact, the average molecular kinetic energy is directly proportional to the temperature. The temperature at which all molecular motion would cease (except for quantum mechanical effects) is termed *absolute zero*. On the Centigrade scale, absolute zero is at $-273\,°C$. This point is used as the zero point of the *absolute* (or *Kelvin*) temperature scale (see Fig. 5.10). The size of the Kelvin degree (°K) is the same as that of the Centigrade degree, and so the temperature of a substance in degrees Kelvin is equal to the temperature in degrees Centigrade plus 273°. Using modern techniques, temperatures approaching 0.000001°K have been produced in the laboratory.

ENERGY

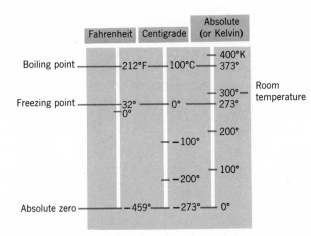

Fig. 5.10 *Comparison of the Fahrenheit, centrigrade, and the absolute (or Kelvin) temperature scales.*

Table **5.2** *Range of Temperatures in the Universe*

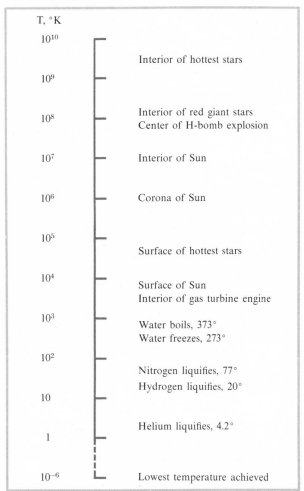

109

It has been found that the volume of certain substances, such as mercury, increases uniformly as the temperature is increased. Therefore, if an amount of mercury is confined within a narrow column in a tube of glass, the level will rise as the temperature is increased and will fall when the temperature is decreased. By marking the points corresponding to the temperatures of ice (0°C) and boiling water (100°C) and dividing the interval into 100 units, we have a conventional mercury *thermometer*. We can now imagine the use of such an instrument in the investigation of various heat phenomena.

HEAT UNITS

Having defined a method for measuring temperature T, we can use this method to arrive at a specification of the amount of heat energy Q that is transferred in any process. Again we use water as the standard substance and say that the amount of heat required to raise the temperature of 1 kg of water by 1°C (from 14.5°C to 15.5°C) is equal to *1 Calorie* (Cal).[4] The Calorie is a unit of energy (or work) and a definite relationship exists between the Calorie and the joule. This conversion factor can be determined by doing a known amount of work on a known mass of water and measuring the resultant increase in temperature. James Prescott Joule performed the first such measurement in 1845 and obtained a value (4140 J/Cal) that is quite close to the presently accepted value

$$1 \text{ Cal} = 4186 \text{ J} \tag{5.21}$$

SPECIFIC HEAT

The quantity of heat required to raise the temperature of an object by an amount ΔT is proportional both to ΔT and to the mass m of the object. Calling the proportionality constant c, we can write

$$Q = cm \, \Delta T \tag{5.22}$$

The quantity c is called the *specific heat* of the substance. The dimensions of c are Cal/kg-°C. The values of the specific heats of a few substances are given in Table 5.3. These values are all less than unity, some substantially less. That is, water (for which $c = 1$, by definition) has an abnormally high specific heat.

Example **5.6**

A 10-g lead bullet is traveling with a velocity of 100 m/sec and strikes a heavy wood block. If, in coming to rest in the block, all of the initial kinetic energy of the bullet is transformed into thermal energy in the bullet, calculate the rise of temperature of the bullet.

The initial kinetic energy is

$$\tfrac{1}{2}mv^2 = \tfrac{1}{2} \times (10^{-2} \text{ kg}) \times (100 \text{ m/sec}) = 50 \text{ J}$$

The temperature rise is

$$\Delta T = \frac{Q}{cm}$$

[4] The Calorie is the unit usually used in specifying the energy content of foods.

Table **5.3** *Specific Heats of Some Materials*

Substance	c (Cal/kg-°C)
air	0.17
aluminum	0.219
copper	0.0932
ethyl alcohol	0.535
gold	0.0316
iron	0.199
lead	0.0310
mercury	0.0333

where $Q = 50$ J. The specific heat of lead is 0.0310 Cal/kg-°C. (see Table 5.3). Therefore,

$$\Delta T = \frac{50 \text{ J}}{(4186 \text{ J/Cal})(0.0310 \text{ Cal/kg-°C}) \times 10^{-2} \text{ kg}} = 38.6°$$

THE SECOND LAW OF THERMODYNAMICS

Since every macroscopic substance has a certain amount of internal energy associated with the motion of its constituent molecules, we can ask the question "Is it possible to extract this internal energy and use it to do work on another substance?" Suppose we have a certain mass m_1 of a material that is at a temperature T_1. If we place this object in contact with another mass m_2 of a material that is at a *lower* temperature T_2, as in Fig. 5.11,

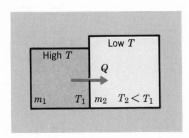

Fig. 5.11 *Heat energy can be transferred from a hotter to a colder body.*

then we know from experience that the pair of objects will eventually come to a common temperature between T_1 and T_2. In other words, some of the internal energy of m_1 has been used to do work on m_2 and has increased the internal energy and temperature of m_2. Heat energy can flow from a hotter body to a colder body. But the reverse is not true. We cannot use the internal energy of m_2 to increase the temperature of m_1 while the temperature of m_2 is *decreased*. Thus, unless work is done by an outside agent, heat energy always flows from objects at higher temperatures to objects at lower temperatures. This is the substance of the *second law of thermodynamics*.

The second law of thermodynamics can be stated in terms of the probability that a system will be in one of the possible configurations that are available to it. For example, if we have a large number of gas molecules in a box, it is most probable that the molecules will be distributed uniformly throughout the volume of the box; it is extremely unlikely that the molecules will be found, at some instant, all in one corner of the box. The latter (unlikely) configuration is one of a high degree of *order*, whereas the uniform distribution of randomly moving molecules constitutes *disorder*. Any ordered system in Nature if left to itself, always tends to proceed spontaneously to a configuration with a lesser degree of order; that is, the trend in natural occurrences is always towards *disorder*.[5] The situation shown in Fig. 5.11 is an example of such a process; the initial configuration is one in which the molecules in m_1 have a high average velocity while those in m_2 have a lower average velocity (that is, $T_1 > T_2$)—thus, there is a certain degree of order in the system. But, when the objects are placed in thermal contact, this order is decreased as the excess energy of the molecules in m_1 is shared with the molecules in m_2. Heat energy always flows in the direction that allows a decrease in the order of a system; in other words, molecular motions always tend to produce a condition that is maximally random, a condition in which the internal energy is shared as equitably as possible among the constituents.

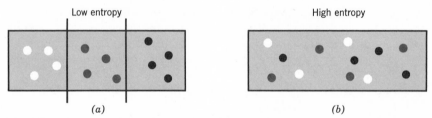

Low entropy   High entropy

(a)   (b)

Fig. 5.12 (a) *An ordered system with a low entropy.* (b) *Removal of the barriers in the box in* (a) *allows the particles to mix; the degree of order is lowered and the entropy is increased.*

ENTROPY

The degree of order in a system can be expressed in a quantitative way by using the concept of *entropy*. An *ordered* system has a *low* entropy; a *disordered* system has a *high* entropy (see Fig. 5.12). The second law of thermodynamics therefore states that in an isolated system some portion of the system may actually experience a decrease in entropy, but such a decrease is always more than compensated by an increase in entropy in the remainder of the system so that the net effect for the system as a whole is an increase in entropy.

Biological systems are not exempt from the second law of thermodynamics. The metabolic processes in a cell increase the order in the cell by forming large molecules from small molecules as, for example, in photosynthesis. Although the entropy of the cell is decreased by these processes, the entropy of the surroundings is increased by an even greater amount.

ENERGY

[5] Any housekeeper will be able to provide ample evidence in support of this statement.

Any isolated system, and indeed the entire Universe, follows a course that continually increases its entropy.

5.10 *Mass and Energy*

$\mathcal{E} = mc^2$

Thus far in our development of physical principles we have discovered several important quantities that remain constant during any physical process. These *conservation laws* relate to the following quantities:

1. Mass   (Section 3.3)
2. Linear momentum   (Section 3.8)
3. Angular momentum   (Section 3.9)
4. Electric charge   (Section 4.3)
5. Energy   (Section 5.5)

Although we have treated mass and energy as being separately conserved, the theory of relativity (Chapter 9) shows that there is an intimate connection between these two quantities and, in fact, that there can be interchanges between them. The mass of a system is related to the total energy of the system in a particularly simple way:

$$\mathcal{E} = mc^2 \qquad\qquad (5.23)$$

This is the famous Einstein-mass-energy relation which shows that it is the *mass-energy* of a system (not the mass nor the energy separately) that is conserved in any physical process.

According to Eq. 5.23 there is a truly enormous amount of energy (*mass energy*) contained in even small amounts of mass. Suppose, for example, that 1 kg of matter could be converted entirely into energy. This process would yield 9×10^{16} J, which is approximately one-half of the amount of energy consumed in the United States each day!

THE LIMITATION ON EXTRACTING USEFUL MASS-ENERGY

Unfortunately, the entire mass-energy of a given substance is not available to be transformed into useful energy; we cannot destroy neutrons and protons—we can only release energy by rearranging them into forms that have different total mass. This is another conservation law: the total number of neutrons and protons in any object remains constant regardless of the processes that it undergoes.

One method of rearranging neutrons and protons within nuclei is the *fission* process in which a heavy nucleus (suitably jostled by a neutron) breaks into two fragments. The difference between the initial mass and the final mass is released as kinetic energy. (However, the total number of neutrons and protons remains the same.)

Example 5.7

How much energy is released in the fission of 1 kg of U^{235}?

The amount of mass-energy that is converted to kinetic energy in the fission process is approximately 200 MeV per nucleus. (This is only about 0.1 percent of the total mass-energy of a uranium nucleus; the other 99.9 percent remains in the masses of the neutrons and protons and is therefore not available for conversion into kinetic energy.) In 1 kg of U^{235} there are approximately 2.5×10^{24} atoms. Therefore, the total energy release is

$$\mathcal{E} = (200 \text{ MeV}) \times (2.5 \times 10^{24}) = 5.0 \times 10^{26} \text{ MeV}$$
$$= (5.0 \times 10^{26} \text{ MeV}) \times (1.6 \times 10^{-13} \text{ J/MeV}) = 8.0 \times 10^{13} \text{ J}$$

We can convert this into another popular unit by noting that the explosion of 1 ton of TNT releases approximately 4.1×10^9 J. Thus, the fission of 1 kg of U^{235} releases an amount of energy

$$\mathcal{E} = \frac{8.0 \times 10^{13} \text{ J}}{4.1 \times 10^9 \text{ J/ton TNT}} \cong 20 \text{ kilotons TNT}$$

This is approximately the size of the original atomic bomb of 1945.

The fission process in any heavy nucleus produces approximately 200 MeV; fission is therefore approximately 0.1 percent efficient in the conversion of total mass-energy into useful energy. On the other hand, chemical burning of fuels such as coal extract only about 10^{-10} of the total mass-energy as useful energy because there is relatively little energy stored in molecules in the form of binding energy. Therefore, fission is approximately 10^7 times more efficient than fuel burning in the generation of energy.

Summary of Important Ideas

Kinetic energy is the result of motion; *potential energy* is the result of work done against an opposing force.

The absolute value of energy has no physical meaning; only *changes* in energy are significant.

The total energy of an isolated system remains constant—energy is *conserved*. Energy may be changed from one form to another (for example, from potential energy to kinetic energy) without loss.

The *internal energy* of a body is that energy due to the motion of the constituent atoms or molecules.

The *first law of thermodynamics* expresses the principle of energy conservation when heat flow and internal energy are included.

The *second law of thermodynamics* expresses the fact that heat cannot be transferred from a colder body to a hotter body without work being done by an outside agent.

A system with a high degree of order has a low value of *entropy*, and conversely. Any physical process that takes place within an isolated system *increases* the entropy of the system.

Mass and energy can be interchanged but the mass-energy of a nucleus can be utilized to do useful work only insofar as protons and neutrons can be rearranged to change the mass of the nucleus; protons and neutrons cannot be destroyed.

Questions

5.1

Kinetic energy is imparted to the blood by the pumping action of the heart. What happens to this kinetic energy?

5.2

One often hears the statement that engines or machines are inefficient, that "they waste energy." Does this mean that the energy is lost? Explain.

5.3

Examine the way in which your body has acquired thermal energy. Trace the history of energy transfer and show that the Sun is the ultimate source of this energy.

5.4

A certain volume of gas contains equal numbers of oxygen and nitrogen molecules. Is there any physical principle that dictates against the molecules arranging themselves with all of the oxygen molecules in one half of the volume and all of the nitrogen molecules in the other half? Explain.

5.5

Water evaporates from a salt solution and leaves behind salt crystals. Crystalline salt is a system with a high degree of order whereas the salt solution is a disordered system in which the atoms have random motion. Has the entropy law been violated? Explain.

5.6

A certain inventor claims to have constructed a machine that will produce 9×10^{19} J of useful energy from a ton of coal. Do you believe his claim? Explain.

Problems

5.1

An amount of work equal to 288 N is expended in pushing a 4-kg block across a frictionless horizontal plane. If the block started from rest, what was the final velocity?

5.2

A man whose mass is 100 kg climbs stairs to a height of 10 m. How much work did he do? Is there a difference between the work required to climb stairs (which are slanted) to a given height and that required to climb a ladder (which is vertical) to the same height? Explain.

5.3

What is the least amount of work that a 100-kg man must do to climb to the top of the Washington Monument (550 feet)? (Actually, the body is not very efficient in such processes and the amount of energy expended is somewhat greater than the calculated minimum.)

5.4

A pile-driver is used to implant a stake in the ground. The resistive force of the ground for the particular stake is 2×10^6 N. The mass of the pile-driver head is 2000 kg and it is lifted to a height of 10 m above the stake. How far is the stake driven at each stroke? (Assume that all of the energy of the pile-driver is used in driving the stake.)

5.5

A water storage tank contains 2000 m³ of water and is at an average height of 40 m above ground level. How much work was required to fill the tank from a reservoir at ground level? How much work can be done by the water if it is piped to a place which is 20 m lower than ground level at the tank site? Is energy conservation violated here? Explain.

5.6

A block of mass 10 kg slides down an inclined plane from a height of 5 m. The length of the plane is 10 m. When the block reaches the bottom it is found to be traveling with a speed of 8 m/sec. How much work was done against the frictional force? What was the average frictional force on the block while it was sliding?

5.7

A 500-kg roller-coaster car starts from rest at a point 30 m above ground level. The car dives down into a valley 4 m above ground level and then climbs to the top of a hill that is 24 m above ground level. What velocity did the car have in the valley and at the top of the hill? (Neglect friction.)

5.8

A 2-kg ball of putty is thrown horizontally with a velocity of 20 m/sec at the rear of a car that is moving away from the thrower with a velocity of 15 m/sec. If the putty ball sticks to the rear of the car upon impact, how much energy is absorbed by the car? Is energy conserved in the process? Explain.

5.9

How much work is required to increase the speed of an automobile ($m = 1500$ kg) from 10 m/sec to 20 m/sec? (Neglect friction.)

5.10

An automobile ($m = 2000$ kg) is traveling on a level road at a speed of 20 m/sec. What is the automobile's kinetic energy? If friction were negligible, could the automobile *coast* to the top of a hill that is 15 m higher than the road? If so, what would be the speed at the top?

5.11

How much work is required to raise a 100-kg object from the surface of the Earth to a height of 1000 miles? If the object were dropped from such a height, what would be its impact velocity (neglecting air resistance)?

5.12

Two small spherical charges of $+3 \times 10^{-4}$ C and -8×10^{-4} C are initially separated by 4 m. What is the change in electrostatic potential energy if the separation is increased by 2 m? How much work has been done? What is the change in electrostatic potential energy if the separation is reduced to 2 m? Is work done on the charges in the latter case?

5.13

A negative helium ion (He$^-$, a helium nucleus with *three* electrons) starts from rest at point A and is accelerated toward a positively charged cylinder B. There is a potential difference of $+10^6$ V between B and A. On reaching B, the helium ion passes through a thin foil that removes all of the electrons and produces a completely ionized helium nucleus (He^{++}). This positive ion is now repelled by the positive charge on B and is accelerated toward C. There is zero potential difference between C and A. How much kinetic energy will the ion have at C? What happens to the electrons? (This is basically the principle of operation of the *Tandem Van de Graaff* accelerator.)

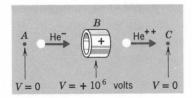

5.14

If a completely ionized carbon nucleus (6 protons and 6 neutrons) moves through a potential difference of 3×10^6 V, what will be its final kinetic energy and final velocity?

5.15

A 1-kg mass of a soft material (such as clay) is dropped from a height of 40 m and sticks to the floor. If $c = 0.2$ Cal/kg-°C for the material, and if all of the heat is produced in the material, what temperature rise will the material experience?

5.16

An amount of work, 10^4 J, is done by hammering on each of two 1-kg bars, one made of copper and the other of gold. Assuming no heat losses, which bar will become hotter and by how much?

5.17

The daily food intake of a certain laborer is rated at 4000 Calories. During a day he performs 10^6 J of work. What fraction of his food intake has been converted into useful work?

5.18

A kilogram of copper shot is placed in a bag. The bag is raised to a height of 10 m and dropped. The process is repeated until 10 drops have been made. Assuming that all of the heat is produced in the copper and that there is no heat loss from the copper, what is the temperature rise of the copper?

5.19

How much U^{235} would have to fission in order to produce an energy release equivalent to 1 Megaton of TNT?

CHAPTER 6
FIELDS

Most of the forces that we encounter in everyday experience are of the *contact* type—we push or pull on something or one object strikes another. To ancient men, contact forces were the only real forces. For the Sun to exert a real force on the Earth seemed impossible because there was no contact between the two bodies. This belief persisted until relatively modern times.

An entirely new concept emerged when Newton established the theory of universal gravitation. According to this theory, the Earth, moon, Sun, and planets all exert forces on one another without contact or any material medium between them through which such forces could be propagated. The term "action at a distance" was used to describe the gravitational interaction (and later, electrical forces as well) because the propagation of the gravitational force through the vacuum of space was inconceivable.

The most popular solution to the dilemma posed by the action-at-a-distance forces was the invention of the *ether*. The ether was believed to be an invisible, weightless jelly—poke it at one point and the pressure induces a strain that can transmit a force to another point.

The ether concept was in vogue until the early 20th century when it was finally laid to rest by Einstein in his theory of relativity. In its place came the *field theory* approach to all action-at-a-distance forces.

6.1 *Scalar and Vector Fields*

A TEMPERATURE FIELD

Consider a thin metal plate that is heated by two flames placed in opposite corners (Fig. 6.1). Heated in this way, the plate will have different temperatures in different places. Imagine now that at some instant of time we measure the temperature of the plate at a large number of positions; we can even imagine that we know the temperature at *every* point in the plate. How do we represent this information? One way is to select a number of temperatures (for example, increments of 10°C) and to draw lines that connect all of the points that have a temperature equal to each of the specified

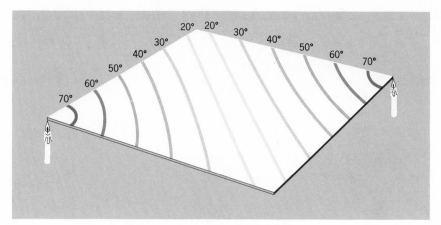

Fig. 6.1 *Temperature variation (in °C) across a thin metal plate heated at two corners.*

values. The resulting picture (Fig. 6.1) gives a representation of the *temperature field* that exists across the plate.

An essential feature of any field is that there is a physical quantity that has a *well-defined value* at every point in the region considered. The equal-temperature lines (*isotherms*) in Fig. 6.1 can therefore never *cross* since such a situation would imply two different values for the temperature at a single point. Furthermore, there must be a smooth variation of the field quantity from point to point within the region. There cannot be a discontinuous change (that is, a sudden jump) in the value of the temperature of the plate from 20°C at one point to 60°C at a neighboring point.

The temperature at a point can be defined by a *single* number. That is, temperature is a *scalar* quantity, and a temperature field is a *scalar field*. *Pressure* is another physical quantity that can be described in terms of a scalar field. Weather maps, on which are marked lines of equal pressure (*isobars*), are actually representations of the scalar pressure field across the country.

VECTOR FIELDS

Many important physical quantities are *vectors* and therefore require not only magnitude but also direction for a complete specification. The flow of water in a river, for example, cannot be represented by a simple scalar field because the *direction* of flow is not included in such a description. Vector quantities must be described by *vector fields*. In the next section we will see how the gravitational and electrical fields (which are vector fields) are specified.

WHEN IS THE FIELD CONCEPT USEFUL?

Some physical quantities, such as the gravitational force vector have a well-defined value at every point in space and field descriptions of these quantities are entirely proper. On the other hand, pressure and temperature, which we have argued can be described by fields, are not quantities that really have a well-defined value at every point. The reason is that these quantities are macroscopic manifestations of large numbers of microscopic effects. It makes no sense to inquire about the pressure or temperature of a single molecule or even a dozen molecules. Large numbers of molecules are required before pressure and temperature have meaning. But the volume occupied by even a million molecules is still small by ordinary standards so that for all practical purposes there is in effect a continuous variation of pressure and temperature from one group of a million molecules to the next. For the description of the gross behavior of a large amount of gas, the field concept is entirely appropriate.

Each particular case must be examined to determine if the physical quantity involved has an acceptable variation in space to make the field description a useful one.

6.2 *The Gravitational Field*

THE FIELD VECTOR

According to Newton's gravitation law, the magnitude of the gravitational force *on* a mass m_2 *due to* a mass m_1 a distance r away is

$$F_{G,21} = G \frac{m_1 m_2}{r^2} \tag{6.1}$$

We know that $\mathbf{F}_G$ is a vector quantity and that the force on m_2 is directed *toward* m_1 (see Fig. 6.2).

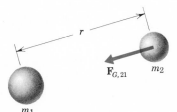

$\mathbf{F}_{G,21}$ m_2

Fig. 6.2 *The source mass m_1 sets up a gravitational field in space. The mass m_2 experiences a force $F_{G,21}$ due to that field.*

m_1

It proves convenient to describe this situation in the following way. The mass m_1 sets up a certain condition in space to which m_2 reacts and m_2 experiences a force directed toward m_1. This "condition" is the *gravitational field* of m_1. (Of course, m_1 also experiences a force directed toward m_2 due to the gravitational field of m_2, but let us continue to consider the effects due to the field of m_1.) Since m_1 in some way produces this gravitational field that attracts m_2, we say that m_1 is the *source* of the field and we refer to m_1 as the *source mass*. Any object (such as m_2) that is placed in this field, at any point, will experience a force that depends on the gravitational field set up by m_1 at that point.

Instead of writing a *force* law specifically for the case of a particular mass m_2, let us divide both sides of Eq. 6.1 by m_2:

$$\frac{F_{G,21}}{m_2} = G \frac{m_1}{r^2} \tag{6.2}$$

6.2
THE
GRAVITA-
TIONAL
FIELD

121

The right-hand side of this expression now depends only on the *distance* of m_2 from m_1 and not on the mass of m_2. That is, the right-hand side is a specification of the *gravitational field* at this distance due to the source mass and will be the same no matter what mass m_2 is placed at this position. Therefore, let us rewrite this expression in a way that emphasizes only the source mass. The new quantity, which is the right-hand side of the equation above and which characterizes the gravitational field of m_1, will be denoted by g:

$$g = G\frac{M}{r^2} \tag{6.3}$$

where the source mass m_1 is now designated by M. The dimensions of g are those of force divided by mass or *acceleration*.

Since the gravitational force $\mathbf{F}_G$ is a vector, the quantity $\mathbf{g}$ is also a vector. The complete specification of the gravitational field (magnitude and direction) due to the source mass M at any point P is given by $\mathbf{g}$, the *gravitational field vector* (see Fig. 6.3).

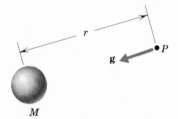

Fig. 6.3 *The source mass M sets up a gravitational field at the point P. This field is specified in magnitude and direction by the field vector* $\mathbf{g}$.

The field vector $\mathbf{g}$ gives the *force per unit mass* on (the *acceleration* of) any object placed in the gravitational field of the source mass M. The gravitational force on a mass m is

$$\mathbf{F}_G = m\mathbf{g} \tag{6.4}$$

This equation is just the vector counterpart of the familiar scalar equation, $F = mg$, that we have used to calculate gravitational forces. In fact, the acceleration due to gravity g, as we have used it, it just the magnitude of the field vector $\mathbf{g}$. Of course, the field vector $\mathbf{g}$ is a more general quantity and varies with position in space, but it does have the magnitude 9.8 m/sec² at the surface of the Earth.

THE PRINCIPLE OF SUPERPOSITION

One of the facts that makes the field concept so useful for the gravitational case (as well as for the electrical case, as we shall see), is that the gravitational force vector and the gravitational field vector obey the *principle of superposition*. That is, if we wish to calculate the gravitational force on a given object due to many other objects (Fig. 6.4), the net force is the vector sum

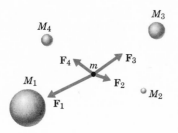

Fig. 6.4 *The net gravitational force on m is the (vector) sum of all the individual forces due to M_1, M_2, M_3, and M_4.*

of all the individual forces; each of these individual forces can be calculated as if the other objects were not present. Thus,

$$\mathbf{F}_{G,\text{net}} = \mathbf{F}_1 + \mathbf{F}_2 + \mathbf{F}_3 + \cdots \tag{6.5}$$

To state that the gravitational force on an object is the vector sum of all contributing forces, each calculated without regard to the others, is not an empty or trivial statement. For example, consider the force on a mass m due to two other masses, M_1 and M_2, as in Fig. 6.5. The principle of

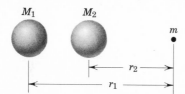

Fig. 6.5 *The force on m due to M_1 is not "screened" by the presence of M_2.*

superposition states that the force on m is the *sum* of the individual forces due to M_1 and M_2. The fact that M_2 lies *between* M_1 and m does not affect the force on m due to M_1. That is, M_2 does not "screen" or "shadow" the force of M_1 on m—it is the same whether M_2 is present or not.

LINES OF FORCE

A diagram or a map of a vector field, such as the gravitational force field of a source mass, is more complex than that of a simple scalar quantity, because both magnitude *and* direction must be specified. Suppose that we begin to map the gravitational force field around a certain source mass M by measuring the force on a small *test mass*.[1] The results of such measurements can be represented by a series of arrows, as in Fig. 6.6. The length of each arrow is proportional to the gravitational force at the end of the arrow (that is, at the end opposite the point) and the direction of the force is given by the direction of the arrow. Alternatively, we can construct around the source mass a set of continuous lines, called *lines of force,* so that at any point the *direction* of the force is the direction of the line of force passing through that point (See Fig. 6.7). The *magnitude* of the force at any point in such a diagram is proportional to the *density* of the lines in the immediate

[1]A "test mass" is an hypothetical object whose mass is so small that it does not disturb the gravitational field of the source mass and whose dimensions are so small that the field of the source mass is essentially uniform throughout the volume of the object.

6.2
THE
GRAVITA-
TIONAL
FIELD

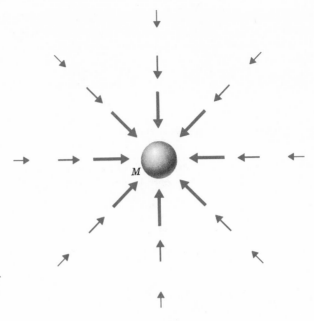

Fig. 6.6 *The mapping of the gravitational field of M. Each arrow represents the force on a test mass placed at the end of the arrow.*

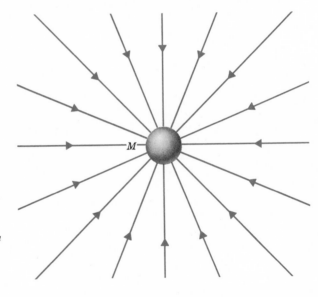

Fig. 6.7 *The lines of gravitational force around a spherical object are all straight lines. The density of lines decreases with the square of the distance from M because the force varies as $1/r^2$.*

vicinity of that point. At a distance r from the center of the mass M, the density of lines of force is proportional to $1/r^2$, as demanded by the radial dependence of the gravitational force law. Therefore, the simple inspection of a lines-of-force diagram reveals where the force is greatest (where the lines bunch together) and where the force is least (where the lines spread out to low density).

Although the lines-of-force scheme is useful in visualizing the force field surrounding an object, it is important to realize that this picture is only an *invention*—there are no rubber-band-like lines that extend through space and exert forces on other objects. Lines of force are not *real*—they serve

FIELDS

only to provide a crutch for our thinking when we consider force-field problems.

For a simple spherical mass, the lines of force are all straight lines in the radial direction. But for objects with complicated shapes or for a group of bodies (even spherical bodies), the lines of force will generally be curved. For example, consider the case of two identical spherical objects that are located close together, as in Fig. 6.8. The lines of force can be plotted by measuring the force on a test mass at many places in the field or by calculating the vector sum of the two gravitational forces at every point.

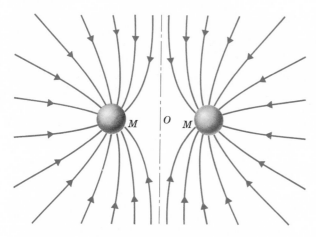

Fig. 6.8 *The lines of force for two identical, nearby objects of mass M. What is the force on a test mass placed on the plane of symmetry (represented by the dot-dash line in the figure)? How will a test mass react in the vicinity of O?*

6.3 *The Electric Field*

THE FIELD VECTOR

Having discussed the gravitational field in detail, it is now a simple matter to apply the same reasoning to the electrical case.

If a test charge q_2 is placed a distance r from a source charge q_1, the force *on q_2 due to q_1* is, according to Eq. 4.3,

$$F_{E,21} = -k\frac{q_1 q_2}{r^2} \tag{6.7}$$

Dividing $F_{E,21}$ by q_2, we obtain a quantity characteristic of q_1:

$$\frac{F_{E,21}}{q_2} = -k\frac{q_1}{r^2}$$

This new quantity, which is the *force per unit charge* is denoted by E and specifies the magnitude of the *electric field* due to q_1. Again, we change the notation and write the source charge as Q. Thus, the electric field of Q (a uniform spherical charge) at a distance r is

$$\boxed{E = -k\frac{Q}{r^2}} \tag{6.8}$$

The quantity that characterizes the electric field is, of course, a vector: **E**. The direction of **E** is arbitrarily chosen to be the direction of the force exerted on a *positive* test charge in the field. Therefore, the field vector of a positive source charge is directed *away* from the source and the field vector of a negative source charge is directed *toward* the source (see Fig. 6.9).

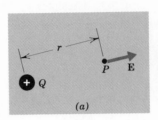 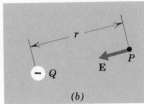

Fig. 6.9 (a) *The electric field vector for a positive source charge is directed* away *from the source.* (b) *The field vector for a negative source charge is directed* toward *the source. The direction of* **E** *is always the direction of the force exerted on a positive test charge in the field.*

The dimensions of **E** are *volts per meter*, V/m:

If we place a test charge q in this electric field, the charge will experience a force given by

$$\mathbf{F}_E = q\mathbf{E} \tag{6.9}$$

The electric field vector obeys the superposition principle:

$$\mathbf{E}_{\text{total}} = \mathbf{E}_1 + \mathbf{E}_2 + \mathbf{E}_3 + \cdots$$

where $\mathbf{E}_1, \mathbf{E}_2, \mathbf{E}_3, \ldots$, are the field vectors at a given point for individual source charges calculated without regard for the other charges.

It is important to realize that the gravitational and electrical fields are *independent* quantities. The two fields can coexist at a particular point in space and neither field influences the other. The total force on a test particle (which possesses both mass and charge) is the vector sum of $\mathbf{F}_G$ and $\mathbf{F}_E$, but it makes no sense to sum the two field vectors, **g** and **E** (the *dimensions* are different). Only the *forces* are the measurable (and therefore the physically significant) quantities.

THE ELECTRIC POTENTIAL

According to Eq. 5.14, the potential energy of a charge q that is a distance r from another charge (now the *source* charge) Q is

$$PE_E = k\frac{Qq}{r} \tag{6.10}$$

If we divide this expression by q, we obtain a quantity characteristic of the source charge Q. This new quantity is called the *electric potential* Φ_E:

$$\Phi_E = \frac{PE_E}{q} = k\frac{Q}{r} \tag{6.11}$$

Φ_E is *potential energy per unit charge* and has dimensions of *joules per coulomb* (J/C), or *coulombs per meter* (C/m), or *volts* (V).

FIELDS

The electric potential for a collection of source charges is just the algebriac sum of the individual potentials; that is, the electric potential obeys the superposition principle:

$$\Phi_{E,\text{total}} = \Phi_1 + \Phi_2 + \Phi_3 + \cdots$$

The work required to move a charge from one point to a second point in the electrostatic field of another charge is the potential energy difference ΔPE_E between the two points. The work *per unit charge* required to make this movement is the change in potential $\Delta \Phi_E$ between the points. We give to this quantity the symbol V:

$$\boxed{\frac{W}{q} = \frac{\Delta PE_E}{q} = \Delta \Phi_E = V} \tag{6.12}$$

V is the *potential difference* or the *voltage* between the two points.

The units of the various electrical quantities that we have introduced are summarized in Table 6.1. A graphical comparison between the gravitational and electrical cases is made in Fig. 6.10.

Table **6.1** *Units of Electrical Quantities*

Quantity	Units
F_E	newtons (N)
Q	coulombs (C)
E	volts/meter (V/m)
Φ_E, V	volts (V)

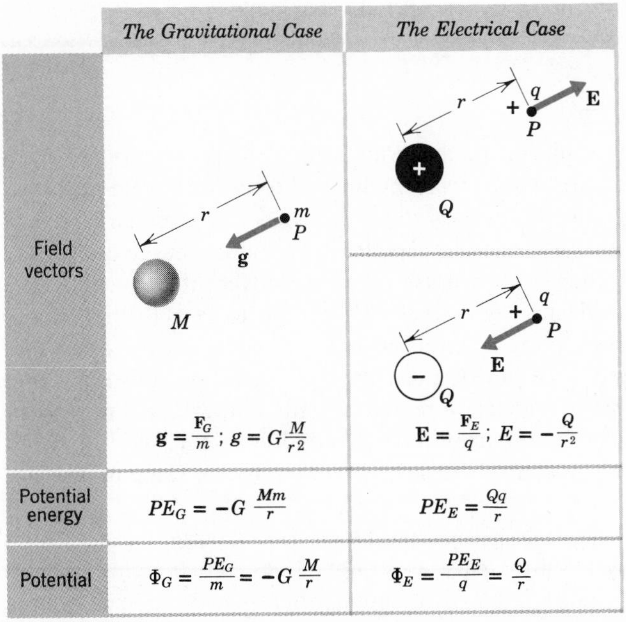

	The Gravitational Case	The Electrical Case
Field vectors	$\mathbf{g} = \frac{\mathbf{F}_G}{m}$; $g = G\frac{M}{r^2}$	$\mathbf{E} = \frac{\mathbf{F}_E}{q}$; $E = -\frac{Q}{r^2}$
Potential energy	$PE_G = -G\frac{Mm}{r}$	$PE_E = \frac{Qq}{r}$
Potential	$\Phi_G = \frac{PE_G}{m} = -G\frac{M}{r}$	$\Phi_E = \frac{PE_E}{q} = \frac{Q}{r}$

Fig. 6.10 *Comparison of the various field quantities for the gravitational and electrical cases.*

Example **6.1**

Compute the electric field and the electric potential at point P midway between two charges, $Q_1 = Q_2 = +5 \times 10^{-6}$ C, separated by 1 m.

Because P is located midway between two identical charges, any test charge placed at this point will experience equal but *oppositely directed* forces due to Q_1 and Q_2. The *net* force is therefore zero, so that $\mathbf{E} = 0$ at P.

Although the electric field at P is zero, this does *not* imply that the electric potential is also zero. The total potential $\Phi_{E,\text{total}}$ is the sum (the *algebraic* sum since potential is a *scalar*) of the potentials due to Q_1 and Q_2:

$$\Phi_{E,1} = k\frac{Q_1}{r_1} = (9 \times 10^9) \times \frac{5 \times 10^{-6}}{0.5} = 9 \times 10^4 \text{ V};$$

$$\Delta_{E,2} = k\frac{Q_2}{r_2} = (9 \times 10^9) \times \frac{5 \times 10^{-6}}{0.5} = 9 \times 10^4 \text{ V}$$

Therefore,

$$\Phi_{E,\text{total}} = \Delta_{E,1} + \Phi_{E,2} = 1.8 \times 10^5 \text{ V}$$

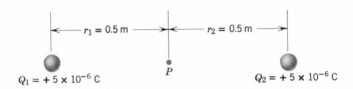

Notice that if either Q_1 or Q_2 is changed from $+5 \times 10^{-6}$ C to -5×10^{-6} C, the electric *potential* will vanish but the electric *field* will not. Therefore, the fact that either the field or the potential is zero in any particular case does not necessarily mean that the other quantity will also be zero; each quantity must be calculated separately.

ELECTRIC FIELD LINES

Because a test mass placed in the gravitational field of a source mass always experiences an attractive force, the lines of force (or the gravitational *field lines*) are always directed *toward* the source mass. Because an electrical test charge will be either attracted or repelled by a source charge depending on the signs of the charges, we must adopt a convention regarding the direction ascribed to electric field lines. We choose to give to the electric field lines for a source charge of either sign the direction of the force that would be experienced by a *positive* test charge. Therefore, the field lines from a positive source charge are directed radially *outward*, whereas those from a negative source charge are directed radially *inward* (see Fig. 6.11). The convention for the electric field lines is therefore the same as that for the electric field vector.

Where do the field lines go? If we had an isolated charge, the field lines would go indefinitely far into space as straight lines. But the concept of a completely isolated charge is an idealization; there exists no such charge in Nature. All macroscopic matter (and, presumably, the entire Universe)

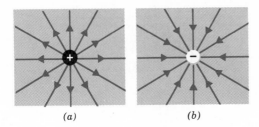

Fig. 6.11 *The electric field lines from a positive source charge (a) are directed radially outward whereas those from a negative source charge (b) are directed radially inward.*

is composed of equal numbers of elementary positive and negative charges and is therefore electrically neutral. (Objects can be charged, but this condition is brought about by separating the positive and negative charge of an originally neutral object.) Let us consider the case of two objects carrying equal and opposite charges (Fig. 6.12). As always, we can generate the field lines by measuring or calculating the magnitude and direction of the force on a positively-charged test body. If we do this, we find that the field lines emerge from the object carrying positive charge and go smoothly in curves that terminate on the negatively-charged object.

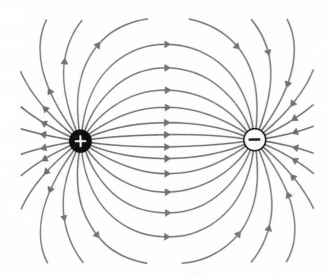

Fig. 6.12 *The electrical field lines originate on the positive charge and terminate on the negative charge.*

If we perform this kind of experiment or do the calculation for an arbitrary assembly of charges (but with zero *net* charge), we always find the same result: *the electric field lines originate on positive charges and terminate on negative charges.* This is one of the important results in the theory of electrostatics.

The conclusion regarding the origination and termination of electric field lines is different from that for gravitational field lines. In the latter case, the field lines have no definite point of origin but extend from infinity to the source mass (at least in our nonrelativistic view).

VOLTAGE

The difference in electric potential between two points is called the *voltage* between the points: $\Delta\Phi_E = V$. Suppose that we have a pair of parallel plates

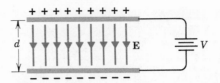

Fig. 6.13 *The magnitude of the electric field between a pair of parallel plates connected to a battery is V/d.*

separated by a distance d and across which there is a voltage V supplied by a battery (Fig. 6.13). The electric field between the plates is uniform and the magnitude is given by the electrical force on a test charge q in the field (Eq. 6.9): $E = F_E/q$. If we move the (positive) test charge from the negative plate to the positive plate, the work required is

$$W = F_E \times d = qE \times d \tag{6.13}$$

or,

$$E = \frac{W}{qd} \tag{6.14}$$

The work done per unit charge, W/q, is just the voltage (Eq. 6.12); therefore, the intensity of the uniform electric field between a pair of parallel plates is

$$\boxed{E = \frac{V}{d}} \quad \text{(uniform field)} \tag{6.15}$$

THE DETERMINATION OF THE ELECTRONIC CHARGE e

In 1911 Robert A. Millikan performed a beautifully simple but highly significant experiment that established the discreteness of the charge on the electron; his experiment also obtained, for the first time, a precise value for this quantity. Millikan set up an electric field between a pair of parallel plates, as in Fig. 6.14. He sprayed a fine mist of oil droplets into this field.

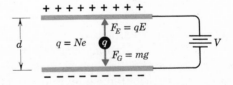

Fig. 6.14 *Schematic diagram of Millikan's oil drop experiment for the determination of e.*

(Oil does not evaporate as readily as water.) Some of the droplets became negatively charged by friction in the process of spraying. Millikan viewed the droplets with a microscope and found that by adjusting the voltage V between the plates to a particular value he could suspend a given droplet in an equilibrium position. When suspended, the downward gravitational force, mg, equaled the upward electrical force, qE, where the magnitude

FIELDS

of E is just V/d. Therefore, at equilibrium[2]

$$mg = qE = \frac{qV}{d}$$

so that

$$q = \frac{mgd}{V} \tag{6.16}$$

The charge on the droplet q is therefore given in terms of measurable quantities. Millikan found that the values of the charge on various droplets, determined in this way, were not of arbitrary sizes. Instead, he found that every charge was an integer number times some basic unit of charge; that is, $q = Ne$, $N = 1, 2, 3, \ldots$. This basic unit of charge is the charge on the electron, $e = 1.6 \times 10^{-19}$ C.

6.5 *The Field of the Nuclear Force*

THE DIFFERENCE BETWEEN THE NUCLEAR FORCE AND $1/r^2$ FORCES

The gravitational field of a spherical object or the electrical field of a spherical charge has an exceedingly simple character. The strength of such a field decreases with distance in accordance with the same geometrical law that governs the variation with distance from the source of the intensity of a quantity, such as light, that is emitted uniformly in all directions into space. This $1/r^2$ variation of the field strength with distance means that a force will be experienced by a test object at any definite distance from the source. It is only at the mathematically well defined but physically unrealizable distance, $r = \infty$, that the force decreases to zero.

The nuclear force, on the other hand, is of a decidedly different character. The effect of this force does not extend to infinity, but is instead confined to extremely small distances around the source. What is the nature of the *nuclear* field? Indeed, is the field concept of any validity in the description of these strange and complicated forces?

HE PION FIELD

The field concept has been applied to the strong nuclear force, but the nature of the problem in this case requires a departure from our previous reasoning regarding fields. The basic new idea that paved the way for our present (and still incomplete) understanding of the nuclear force was provided by the Japanese physicist Hideki Yukawa (1907–   ) in 1935. Yukawa hypothesized that two nucleons experience an attractive force at small distances because of the exchange between them of a new elementary particle which had not yet been observed at the time) called a *meson*. A meson is a particle with mass intermediate between the mass of an electron and that of a proton. Several brands of mesons are now known, but the

---

[2] Millikan actually used a *dynamic* instead of a *static* method for determining e in his oil drop experiment, but the distinction is not important here.

6.5
THE FIELD
OF THE
NUCLEAR
FORCE

131

meson that is responsible for the strong nuclear force has been found to have a mass approximately 273 times that of an electron (or about 0.15 of the mass of a nucleon) and is called the π *meson* or *pion*.

The *pion exchange force*, when treated in detail, requires complicated mathematics, but a qualitative description of an exchange force can be made. Suppose that two boys are grappling for control of a basketball. One boy grabs the ball from the other boy and then the second boy snatches it back again; the process is repeated over and over. This continual exchange of the ball results in each boy being pulled toward the other; that is, there is an attractive "basketball exchange force."

Suppose that a proton and a neutron collide. The collision process is controlled by the nuclear force (the pion exchange force) that acts between the two particles. Where does the pion that mediates the interaction come from? In a sense, the pion was always "there." It is very useful to picture the proton as consisting of a neutron and a positively-charged pion:

$$p \longleftrightarrow n + \pi^+$$

That is, a proton is equivalent to a neutron "core" surrounded by a "meson cloud." Or, in the field picture, we say that there is a pion field around the neutron and that the combination appears as a proton. Therefore, in the collision of a neutron and a proton, the π^+ meson is exchanged between the two neutrons and mediates the nuclear force between the particles. This exchange of the pion is an extremely rapid process, requiring only about 10^{-23} sec.

If the proton and the neutron differ in mass by only about 0.1 percent, how can the proton be considered to be a neutron combined with a pion when the latter has a mass of 15 percent of the proton? We cannot answer this question in detail until we have discussed the famous *uncertainty principle* of quantum mechanics (Section 10.5). In essence, this principle states that we can conceive of processes, such as $p \rightarrow n + \pi^+$, in which the mass does not balance, as long as the system reverts to its original state, that is, $n + \pi^+ \rightarrow p$, within a sufficiently short period of time. Since the complete exchange process takes only about 10^{-23} sec, the exchange of a pion between the colliding neutron and proton is allowed.

By using the pion field concept of the strong nuclear force, a great deal of progress has been made, although we still have much to learn before we can claim to understand completely this basic force of Nature. Presumably, there is some undiscovered elementary particle that mediates the weak force between electrons and neutrons, but as yet we have no clear conception of the nature of this particle.

6.6 *Energy in the Field*

DOES AN OBJECT POSSESS POTENTIAL ENERGY?

When we raise an object to a height h above the surface of the Earth, we say that the object possesses a gravitational potential energy mgh relative to its initial position. But does the *object* really possess this potential energy? Or does the *Earth* share in the energy? According to our field description of the gravitational interaction we should not ascribe the increase in potential energy to *either* body. An amount of work *mgh* has been done *on the field* by changing the relative positions of the two bodies and it is the *gravitational field* that has acquired this energy. The energy can be recovered from the field by allowing the field to set the objects into motion. Similar comments also apply for the electrical and nuclear force fields.

Although it is well to keep in mind that the *field* possesses potential energy, we shall usually follow our previous custom and continue to refer to the potential energy "of a body" or the potential energy "or a system of bodies."

DOES THE FIELD REALLY POSSESS ENERGY?

We have argued that a field can possess energy. But is this just a fiction? Can a vacuum, completely devoid of any material particle, actually retain a physical quantity as real as energy? Consider the following situation. A space vehicle has landed on the moon and is telemetering information back to the Earth via radio waves (Fig. 6.15). A radio wave, as we shall see in

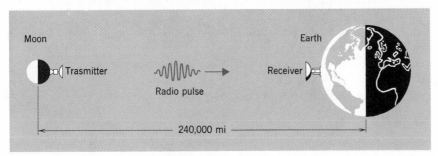

Fig. 6.15 *It requires approximately 1.3 sec for a radio signal to be transmitted from the moon to the Earth. During this interval, the transmitted energy resides in the electromagnetic field.*

Chapter 8, is a propagating distrubance in an electromagnetic field. The transmitter on the moon sends out a pulse that contains a certain amount of energy. This pulse propagates with the velocity of light and about 1.3 sec later it can be detected by a receiver on the Earth. Energy has been transmitted from an instrument on the moon and detected by another instrument on the Earth. Where was the energy during the 1.3-sec interval between transmission and reception? The energy conservation principle insures that the energy was *somewhere;* it could only have been contained in the *electromagnetic field.*

Summary of Important Ideas

In the broadest sense, any physical quantity that is smoothly varying and well-defined at all points in space can be considered a *field* quantity.

Field quantities can be either *scalars* or *vectors.*

The *gravitational field quantity* is the force per unit mass (that is, the *acceleration*) acting on a test mass.

Gravitational field lines are always directed *toward* the source mass.

The *electric field quantity* is the force per unit charge acting on a test charge.

Gravitational (electrical) *potential* is the potential energy per unit mass (charge).

Electric field lines always *originate* on positive charges and *terminate* on negative charges.

The electric field between a pair of parallel, uniformly charged plates is *uniform*.

Fields can possess *energy*. There are *three* basic types of energy: kinetic energy, field energy, and mass energy.

Questions

6.1

Under what conditions can the following be usefully considered to be fields? Specify whether the field is *scalar* or *vector*.
(a) The mass density distribution in the Earth.
(b) The population density in a country.
(c) The population density in a city block.
(d) The density of stars in a galaxy.
(e) The flow of air masses in the atmosphere.

6.2

In a certain oval-shaped pool, the water inlet is at the bottom near one end of the pool. The outlet is at the bottom near the opposite end. Sketch flow diagrams that represent water flow in this pool as seen from above and from one side. What kind of field is this?

6.3

If a body is released in a force field to which it is sensitive, its motion does not necessarily follow a line of force. Explain why. Under what special conditions would the motion be exactly along a line of force?

6.4

Two concentric, hollow spheres carry equal and opposite charges. Argue whether an electric field exists in each of the regions *A, B,* and *C.* Reexamine the situation if both spheres carry equal charges of the same sign.

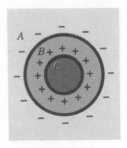

Problems

6.1

What is the magnitude of the gravitational field vector at the surface of the Earth?

6.2

What is the difference in the gravitational potential energy of a 2-kg mass between the base and the top of the Washington Monument ($h = 555$ ft)?

6.3

A charge $Q = -5 \times 10^{-7}$ C is located at the origin of a coordinate system. Find the electric field vector and the potential at the point $x = 4$ m, $y = 4$ m.

6.4

At a certain point P in space a source charge produces an electric field of 300 V/m in the $+x$ direction. At this same point another source charge produces a field of 600 V/m in the $+y$ direction. What force will a proton experience at P?

6.5

An electron is placed in a uniform field of 10^5 V/m. What force does the electron experience?

6.6

Consider the proton to be a uniformly charged sphere with a radius of 10^{-15} m. (This is a dubious model of the proton.) What is the electric field at the surface of the proton?

6.7

A sphere of radius 1 m carries a uniform surface charge of 10^{-6} C/m^2. What is the electric field and the potential at the surface of the sphere?

6.8

Two points, A and B, have electric potentials of $+100$ V and -150 V, respectively. How much work is required to move an electron from A to B? Is the same amount of work required to move a proton from A to B?

6.9

Two parallel plates are separated by 2 cm. A battery is used to put a potential difference of 600 V across the plates. What electrical force will an oil droplet carrying a charge of $4e$ experience in the field between the plates?

CHAPTER 7
ELECTRIC CHARGES
IN MOTION

The motion of electric charges is a fundamental aspect of much that takes place in the Universe. Moving charges influence the dynamics of the Sun, provide the Earth with its magnetism, and cause spectacular auroral displays. And, of course, moving charges in the form of electric current perform countless tasks in our present-day electrically-oriented world by operating all manner of electrical motors, radio circuits, and other devices. There is, in fact, almost no area of modern society that does not depend, in one way or another, on electromagnetic effects that are attributable to the motion of electric charges.

7.1 *Electric Current*

THE TRANSPORT OF ELECTRONS

The movement of electric charge from one position to another constitutes a *current*. When an electric current flows in a wire the positively-charged nuclei of atoms remain essentially stationary compared to the movement of the much less massive electrons. Thus, in almost all of the cases with which we shall be concerned, electric current is due to the motion of *electrons*.

In any material that conducts a current (whether that material is in solid, liquid, or gaseous form), a fraction of the electrons are not attached to specific atoms but are free to move about in the material. In the absence of an applied electric field, these *free electrons* (or *conduction electrons*) move at random, colliding with the stationary atoms at frequent intervals and thereby changing their directions of motion. Across any given surface in the material, just as many electrons will move in one direction as in the other (Fig. 7.1). Therefore, there is no *net* transport of electrons across the surface and the electric current is *zero*.

If the ends of the conductor are attached to the terminals of a battery (Fig. 7.2), an electric field is established within the conductor. There will now be a net transport of electrons across any surface that is perpendicular to the field lines. That is, in Fig. 7.2, there will be a few more electrons crossing the surface from *right* to *left* than will be crossing from *left* to *right*.

Fig. 7.1 *If no electric field is present, the motion of the free electrons is random.*

The direction in which the electrons drift in an electric field is opposite to the direction of the field lines because electrons carry a negative charge. However, just as we arbitrarily elected to consider the direction of electric field lines to be the direction of the force on a *positive* charge, it is traditional to refer to the direction of *current* flow as the direction in which *positive* charges would move (even though it is only the electrons that actually move).

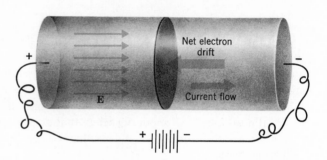

Fig. 7.2 *The application of an electric field to a conductor causes a net motion of the electrons in the direction opposite to the electric field lines. By convention, the current is said to flow in the same direction as the field lines.*

A negative current (or an electron current) in one direction would produce exactly the same effects and is therefore equivalent to a positive current in the opposite direction. We shall always use the terms *current* or *current flow* to mean this so-called *conventional* current which flows in the same direction as the electric field lines. The terms *electron flow* or *electron current* will be used to refer explicitly to the motion of the electrons. The advantage in retaining this seemingly awkward convention will soon become apparent.

THE DEFINITION OF CURRENT

Electric current is defined quantitatively as the *rate* at which electric charge flows across a surface. When 1 C of positive charge crosses a given surface in 1 sec, a current of 1 *ampere* (A) is flowing. In general, the current I is given by the net charge Q crossing a surface during a time t:

$$I = \frac{Q}{t} \tag{7.1}$$

The unit of current is

$$1 \text{ A} = 1 \text{ C/sec} \tag{7.2}$$

Of course, if we wish to define the *total* current flow in a wire, the surface must be one that is a complete cross section of the wire.

ELECTRIC
CHARGES
IN MOTION

Household circuits are usually capable of carrying currents of 15–20 A and household appliances usually require currents of a few amperes. A lightning stroke may involve a current of 100,000 A, whereas the current flowing through a camera's light meter may be only $1\mu A = 10^{-6}$ A.

ELECTRON DRIFT VELOCITIES

With what net velocity do electrons move along a current-carrying wire? A typical type of household copper wire has a cross-sectional area of approximately 1 mm^2. A 1-mm length of this wire has a volume of 1 mm^3. Each cubic millimeter of copper contains approximately 8×10^{19} atoms and each atom contributes one free electron. If the electrons move with a net velocity (or *drift* velocity) of 1 mm/sec, all of the free electrons in the segment will pass a point in the wire in 1 sec. Hence, the current is $I = Q/t = (8 \times 10^9$ electrons$) \times (1.6 \times 10^{-19}$ C/electron$)/1$ sec $\cong 13A$, which is about the current limit for this type of wire. Thus, we see that because conductors contain such enormous numbers of free electrons, only very small drift velocities are necessary to produce sizable currents. It should be remembered that this net drift velocity of a mm/sec is superimposed on the large random electron velocities of $\sim 10^6$ m/sec.

7.2   *Electrical Resistance and Electrical Power*

OHM'S LAW

When a wire is connected to a battery (as in Fig. 7.2), how much current will flow? Many materials have the property that the current flow through a particular sample is directly proportional to the voltage across the sample. That is, if a voltage of 6 V causes a current of 2 A to flow, a voltage of 12 V will cause a current of 4 A to flow, and so forth. We can represent the relationship between the voltage and the current by

$$I = \frac{V}{R} \text{ or } V = IR \qquad\qquad (7.3)$$

where the proportionality constant R is called the *resistance* of the sample. The higher the resistance, the smaller the current for a given voltage. This relationship is called *Ohm's law*, in honor of its discoverer, George Simon Ohm (1787–1854). If a voltage of 1 V causes a current of 1 A to flow through a certain sample, the sample is said to have a resistance of 1 ohm (1 Ω).

Copper is one of the best conductors known (at room temperature); only silver is better, and only slightly so. A 100-m length of household copper wire, with a cross-sectional area of 1 mm^2, has a resistance of about 1.8 Ω. The resistance of a similar piece of silver is 1.6 Ω, and for carbon the value is 3500 Ω. The resistance of an ordinary household light bulb (the filament of which consists of fine tungsten wire) is about 100 Ω.

Example **7.1**

An electrical automobile starter requires 8 A to operate. What is the resistance of the starter?

7.2
ELECTRICAL
RESISTANCE
AND
ELECTRICAL
POWER

An automobile battery has a voltage of 12 V; therefore, solving Eq. 7.3 for R,

$$R = \frac{V}{I} = \frac{12 \text{ V}}{8 \text{ A}} = 1.5 \ \Omega$$

ELECTRICAL POWER AND ENERGY

Power is the rate at which work is done (or, equivalently, the rate at which energy is used). If an amount of work W is done in a time t, the average power expended during that time is

$$P = \frac{W}{t} \tag{7.4}$$

The unit of work (or energy) is the *joule*, so the unit of power is the J/sec, which we call the *watt* (W):

$$1 \text{ watt} = \frac{1 \text{ joule}}{\text{second}}; \qquad 1 \text{ W} = 1 \text{ J/sec} \tag{7.5}$$

Suppose that a voltage of 1 V causes a current of 1 A to flow in a certain piece of wire. The product of voltage and current is

$$(1 \text{ V}) \times (1 \text{ A}) = \left(1 \frac{\text{J}}{\text{C}}\right) \times \left(\frac{1 \text{ C}}{\text{sec}}\right) = 1 \frac{\text{J}}{\text{sec}} = 1 \text{ W} \tag{7.6}$$

where we have used the definitions of the volt (1 J/C) and the ampere (1 C/sec). We see that this product is equal to 1 W; that is, the rate at which energy (electrical energy) is being expended in the wire is 1 W. In general, for a voltage V and a current I, the power is

$$\boxed{P = VI} \tag{7.7}$$

and using Ohm's law, $V = IR$, to substitute for V, we can express the power supplied to a resistance R by a current I as

$$\boxed{P = I^2 R} \tag{7.8}$$

The total amount of energy consumed by an element of an electrical circuit is equal to the *rate* of energy use (that is, the *power*) multiplied by the total time:

$$W = Pt$$

If a current of 2 A flows through a 5-Ω resistance for 1 hr, the energy used is

$$W = Pt = I^2 R \times t$$
$$= (2 \text{ A})^2 \times (5 \ \Omega) \times (3600 \text{ sec}) = 72{,}000 \text{ J}$$

In electrical practice we usually express energy in units of watts multiplied by the time in hours. For the case above,

$W = (2 \text{ A})^2 \times (5 \ \Omega) \times (1 \text{ hr})$

$\quad = 20 \text{ watt-hours (Wh)}$

$\quad = 0.020 \text{ kilowatt-hours (kWh)}$

The kWh is the usual unit by which electrical energy is sold by commercial power companies. A typical household user pays a few cents per kWh.

Example **7.2**

A room heater has a resistance of 10 Ω and is used continuously for one day. What is the cost of this operation if the billing rate is 2 cents per kWh?

Ordinary household voltage in 110 V. Therefore, the current drawn by the heater is

$$I = \frac{V}{R} = \frac{110}{10} = 11 \text{ A}$$

The power is

$$P = I^2 R = (11 \text{ A})^2 \times (10 \ \Omega) = 1210 \text{ W}$$

The energy used in one day is

$$W = Pt = (1210 \text{ W}) \times (24 \text{ hr}) = 29 \text{ kWh}$$

and the cost is

$$\text{Cost} = (29 \text{ kWh}) \times (\$0.02/\text{kWh}) = \$0.58$$

The resistance to current flow in a material is caused by the free electrons colliding with the atoms of the material and thereby causing them to be agitated more severely than in their normal state. The flow of current therefore raises the internal energy of the material, and electrical energy is converted into thermal energy. When current flows through the filament of a light bulb, some of the energy used appears as light, but the efficiency of a light bulb is very low and most of the energy—about 98 percent— appears as heat. Indeed, in almost all ordinary applications of electricity, most of the electrical energy is consumed in heating effects.

7.3 *Magnetism*

THE MAGNETIC COMPASS

Magnetism, in the form of magnetic rocks (or *lodestones*), was discovered in ancient times. From this discovery, the *magnetic compass* was developed, and the compass provides us with the standard by which we define directions in describing magnetic effects. The end of a compass magnet which is *north-seeking* is called the *north pole* or *N pole* of the magnet. Similarly, the south-seeking end is the south or S pole of the magnet.

Magnets have the familiar property (similar to that of electric charges) that *opposite poles attract* and *like poles repel* (Fig. 7.3). Therefore, the north-seeking N pole of a compass magnet actually is attracted to and points toward the Earth's S magnetic pole which is located near (but not at) the geographic north pole (see Fig. 7.4).

Fig. 7.3 *Like magnetic poles repel and opposite magnetic poles attract.*

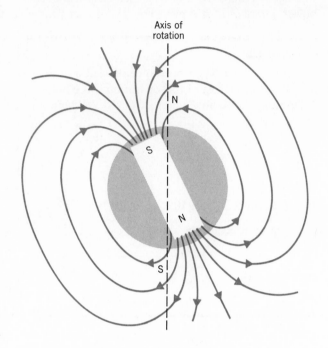

Fig. 7.4 *The Earth's magnetic field is like that of a bar magnet with the S pole near the north geographic pole.*

MAGNETIC FIELDS

We know that at every point in the vicinity of the Earth a compass will assume a definite direction. That is, there is a *magnetic field* due to the Earth's magnetism just as there is a gravitational field due to the Earth's mass. The magnetic field of the Earth or that of any magnetic material, can be mapped by using a compass. (This instrument will indicate the *direction* of the field at any point but it will not directly provide information regarding the *strength* of the field.) Compass measurements of the field of a simple bar magnet show that the magnetic field lines are as indicated in Fig. 7.5. By convention, we take the direction of the field lines to be the direction in which the N pole of a compass magnet points: that is, the field lines in the region external to the bar magnet run *from the N pole to the S pole,* as in Fig. 7.5. Similar measurements of the Earth's magnetic field show that the Earth's magnetism is practically the same as that of a bar magnet (Fig. 7.4).

ELEMENTARY MAGNETS

If a bar magnet is cut into two pieces, as in Fig. 7.6, we find that the two halves are themselves complete magnets with N and S poles in the same orientation as the original magnet. Further division of the magnet produces additional magnets, again with N and S poles oriented in the same direction as the original magnet.

ELECTRIC
CHARGES
IN MOTION

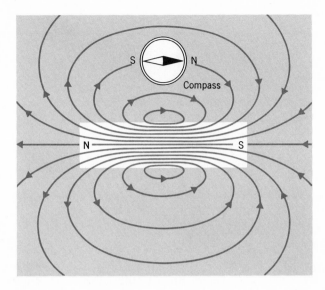

Fig. 7.5 *Magnetic field lines for a simple bar magnet.*

Fig. 7.6 *Cutting a magnet produces two magnets with N and S poles in the same orientation as the original magnet.*

What will happen if we continue this process of division down to the atomic level? Can we ever separate the N pole from the S pole? As we shall discuss in the next section, even individual atoms can behave as microscopic but *complete* magnets with N and S poles. In fact, because we cannot take an atom apart and retain in its components the complete magnetic effects of the atom as a whole, we must conclude that the N and S poles of a magnet have no independent existence.

7.4 *Magnetic Fields Produced by Electric Currents*

THE FIELD OF A CURRENT-CARRYING WIRE

The magnetism of a bar magnet appears to be a completely static affair. The magnetic material is electrically neutral and there does not appear to be any electrical current flowing in the magnet. What, then, is the connection between electrical currents and magnetism? Until early in the 19th century, electricity and magnetism were thought to be two independent phenomena. In 1820 Hans Christian Oersted (1777–1851), a Danish physicist, discovered (quite by accident) that a current-carrying wire influenced the orientation of a nearby compass magnet. The reason for the phenomenon that Oersted observed is that a current-carrying wire produces a magnetic field which can influence the orientation of a compass placed in the vicinity of the wire. Measurements with a compass show that the field lines near a current-

7.4
MAGNETIC
FIELDS
PRODUCED BY
ELECTRIC
CURRENTS

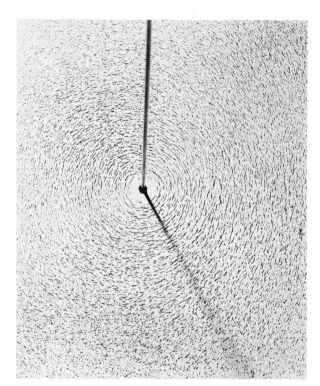

Fig. 7.7 *The circular magnetic field surrounding a current-carrying wire is made "visible" by sprinkling iron filings on a sheet of paper whose plane is perpendicular to the wire. The tiny pieces of iron take up positions with their long dimensions along the field lines.*

carrying wire are *circles* centered on and perpendicular to the wire, as shown in Fig. 7.7.

RIGHT-HAND RULE

We can establish the *direction* of the magnetic field lines due to a current-carrying wire by observing the orientation of a compass magnet when placed in the vicinity of the wire. The results of such an experiment are summarized in the so-called *right-hand rule:*

If a current-carrying wire is grasped with the right hand in such a way that the thumb is in the direction of conventional current flow, then the fingers encircle the wire in the same direction as the magnetic field lines (see Fig. 7.8).

The magnetic field vector (analogous to the electric field vector **E**) is given the symbol **B**.

We see here one of the advantages of having adopted the convention of always using the term *current flow* to mean the (equivalent) flow of *positive* charge, namely, that it permits us to use a *right*-hand rule for the direction of the field lines and the field vector **B**. We have already defined the direction of the angular momentum vector (Section 3.10) in terms of the direction of advance of a right-hand screw and we shall later have additional conventions regarding *right* hands. Therefore our choice allows us to specify the directions of these vector quantities in terms of rules always using the *right* hand.

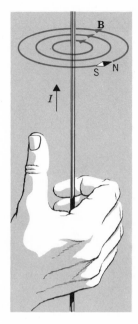

Fig. 7.8 *Illustration of the right-hand rule for determining the direction of the magnetic field lines due to a current flowing in a wire.*

ELEMENTARY MAGNETS

The smallest units of magnetism are found at the atomic level. In a highly simplified model of atomic structure, electrons are considered to move around the atomic nucleus in definite orbits. A single electron that executes a circular orbit around a stationary positively-charged nucleus is shown in Fig. 7.9. The motion of this electron is equivalent to a current loop (but with the current flowing in the direction opposite to the electron velocity). By using the right-hand rule we can determine the direction of the magnetic field in the vicinity of the moving charge. But because the electron orbit is a loop, the field lines are bunched together inside the loop and are spread out in the region outside the loop. The net result is a field that is similar to that of a bar magnet (Fig. 7.9). Indeed, the magnetic field of a bar magnet is just the sum of a large number of elementary atomic fields.

WHERE DO MAGNETIC FIELD LINES BEGIN AND END?

We have previously seen (Section 6.4) that electric field lines originate on positive charges and terminate on negative charges. The lines that specify the magnetic field are distinctly different in character: *magnetic field lines have no beginning and no end*. Thus, the field lines surrounding a straight current-carrying wire are circles. Even the field lines of a bar magnet or electromagnet do not begin at the N pole and end at the S pole; the lines extend through the interior of the bar or core without termination.

The fact that magnetic field lines have neither beginning nor end is equivalent to the statement that isolated magnetic poles (*monopoles*) do not exist. Electric field lines originate and terminate on electric monopoles (charges) but there are no magnetic monopoles to terminate the magnetic field lines. The continuity of magnetic field lines and the nonexistence of magnetic monopoles is one of the important facts of electromagnetism.

7.4
MAGNETIC
FIELDS
PRODUCED BY
ELECTRIC
CURRENTS

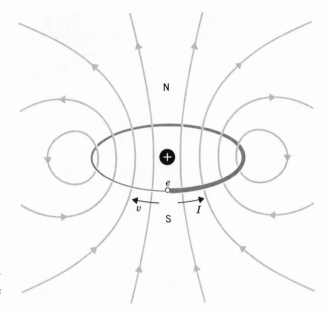

Fig. 7.9 *An electron moves around a stationary atomic nucleus and produces a magnetic field equivalent to that of a current loop. Notice that the direction of current flow is opposite to that of the electron's velocity because the electron carries a negative charge.*

7.5 *Effects of Magnetic Fields on Moving Charges*

THE STRENGTH OF THE MAGNETIC FIELD

Although we have completely specified the *direction* of the magnetic field lines (or, equivalently, the direction of the magnetic field vector **B**), we have as yet made no quantitative statement regarding the *strength* of the field (that is, the *magnitude* of **B**). In the case of the electric field, the magnitude of **E** was defined in terms of the *force* on a stationary test charge in the field. Similarly, we can define the magnitude of **B** in terms of the force exerted by the field on a test charge. But a test charge that is *stationary* in a magnetic field experiences no force. Only in the event that the test charge is in motion is there a magnetic force on the charge.

In a given magnetic field it is found that the magnetic force is directly proportional to both the charge and the velocity of the test particle. The proportionality factor that connects the magnetic force F_M with the charge and velocity (in units of c, the velocity of light) of the test particle is the magnetic field strength B:

$$F_M = qvB \qquad \text{(for } \mathbf{v} \perp \mathbf{B}) \tag{7.9}$$

This equation shows that the magnetic force on a moving charged particle is *directly proportional* to the charge on the particle, the velocity of the particle, and the strength of the magnetic field.

The dimensions of B are those of (*force*)/(*charge* × *velocity*), to which we give the special name, tesla (T). Some typical values of B as found in the Universe are shown in Table 7.1 (Another unit commonly used to measure magnetic field strength is the *gauss* (G): $1 \text{ T} = 10^4 \text{ G}$.)

ELECTRIC
CHARGES
IN MOTION

146

Table **7.1** *Range of Magnetic Field Strengths in the Universe*

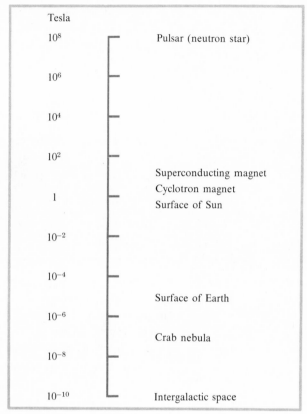

Tesla		
10^8		Pulsar (neutron star)
10^6		
10^4		
10^2		Superconducting magnet
1		Cyclotron magnet
		Surface of Sun
10^{-2}		
10^{-4}		
		Surface of Earth
10^{-6}		
		Crab nebula
10^{-8}		
10^{-10}		Intergalactic space

THE DIRECTION OF THE MAGNETIC FORCE

The quantities v and B that appear on the right-hand side of Eq. 7.9 are the magnitudes of the vectors **v** and **B**. And, of course, F_M is the magnitude of the force vector $\mathbf{F}_M$. What is the relationship among the directions of these three vectors?

The case of the magnetic force is distinctly different from that of the electric force. As shown in Fig. 7.10, the electric force vector $\mathbf{F}_E$ that acts

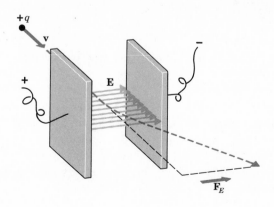

Fig. 7.10 *The electric force vector* $\mathbf{F}_E$ *has the same direction as the electric field vector* **E**.

147

on a positive test charge has the *same* direction as the electric field vector
E and is independent of the direction of the velocity vector **v**. On the other
hand, it has been found experimentally that when a charged particle enters
a magnetic field, the direction of the magnetic force is *perpendicular* to both
v and **B** (Fig. 7.11).

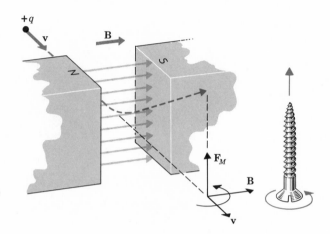

Fig. 7.11 *The magnetic force is perpendicular to
both **v** and **B**. Application of the right-hand rule
shows that the direction of* **F**$_M$ *is up.*

The direction of **F**$_M$ relative to **v** and **B** is given by another right-hand
rule: The vector **F**$_M$ has the same direction as that of the advance of a
right-hand screw when rotated in the sense that moves the vector **v** toward
the vector **B** (see Fig. 7.12*a*). Alternatively, we can state the rule in the

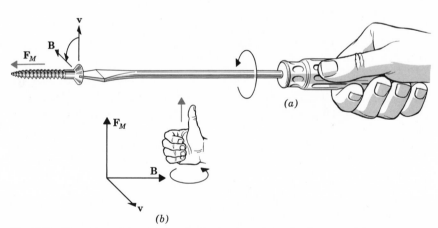

Fig. 7.12 *Illustration of the right-hand rules for determining the direction of the magnetic
force* **F**$_M$ *on a moving positive charge.*

following way: Point the fingers of your right hand in the direction of **v**
and curl the fingers toward the direction of **B**; the thumb then points in
the direction of **F**$_M$ (see Fig. 7.12*b*). Note that the rules apply to the case
of a particle carrying a *positive* charge; for a negatively-charged particle
the direction is *opposite* to that given by the rules.

ELECTRIC
CHARGES
IN MOTION

148

The maximum force exerted on a moving charged particle by a magnetic field occurs when the velocity vector of the particle is perpendicular to the field vector **B**, as in Fig. 7.11. The magnitude of this maximum force is given by Eq. 7.10. If **v** is not perpendicular to **B**, the force proportional to the component of **v** perpendicular to **B** (i.e., to $v_\perp$)

$$F_M = qv_\perp B \qquad (7.10)$$

When **v** is parallel to **B**, the force is zero (because $v_\perp = 0$).

MAGNETIC FIELDS AND RELATIVE MOTION

Suppose that we have a charge q and a meter that is sensitive to a magnetic field; suppose that both are at rest in some coordinate system, as in Fig. 7.13a. Clearly, the meter will show zero field, $B = 0$. (There is, of course, an *electric* field at the position of the meter.) However, if the charge is in motion, as in Fig. 7.13b, we know that this is equivalent to a current and that there is produced a magnetic field which will be registered by the meter. But it is only the *relative* motion of the charge and the meter that is important. (The laws of physics are the same in all inertial reference frames.) Therefore, if q remains at rest in the coordinate system and the *meter* moves, as in Fig. 7.13c, this is entirely equivalent to the situation in Fig. 7.13b, and the presence of a magnetic field will again be shown by the meter. *A magnetic field is produced only by a changing or moving electric charge.*

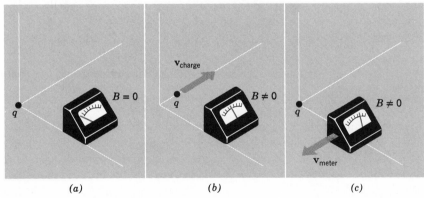

Fig. 7.13 *A magnetic field is produced solely as the result of relative motion between the charge* q *and the "**B** meter."*

7.6 *Orbits of Charged Particles in Magnetic Fields*

CIRCULAR ORBITS

A charged particle can be started into motion by allowing it to be accelerated by an electric field. Once in motion, if we project the charged particle into a uniform magnetic field, we find that the magnetic force exerted on the particle by the field causes it to move in a *circular* path. (We must assume here that the particle moves in a vacuum; otherwise, the particle will collide

7.6
ORBITS OF
CHARGED
PARTICLES IN
MAGNETIC
FIELDS

with and lose energy to other particles and the orbit will not be circular—see Fig. 7.14.)

Consider a uniform magnetic field **B** and a charged particle of mass m moving in the field with a velocity **v**, where $v \perp B$, as in Fig. 7.15. The magnetic field exerts on the particle a force of constant magnitude:

$$F_M = qvB$$

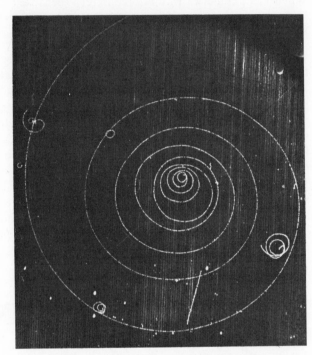

Fig. 7.14 *The path of a charged particle in a bubble chamber (consisting of liquid hydrogen) is made visible by the many tiny bubbles that are formed in the wake of the particle. This photograph shows the orbit of a fast electron in a bubble chamber which is in a strong magnetic field. The electron loses energy through collisions with the hydrogen atoms and so the radius of the orbit decreases, causing the electron to move in a spiral path. The tracks of some secondary electrons released in encounters with hydrogen atoms can be seen near the main track.*

The direction of this magnetic force is always perpendicular to the instantaneous direction of motion of the charged particle, that is, $F_M \perp v$. Hence, there is no component of the force in the direction of motion and *no work is done on the particle by the magnetic field*. Although the direction of motion is continually changing as a result of the magnetic force, the *speed* of the particle (that is v) is constant and the kinetic energy remains always the same.

The magnetic force produces a centripetal acceleration of constant magnitude that is always perpendicular to **v**:

$$a_c = \frac{F_M}{m} = \frac{qvB}{m} \tag{7.11}$$

Thus, the particle moves in a *circular* orbit. In terms of the velocity and the orbit radius, the centripetal acceleration is given by (see Eq. 2.24)

$$a_c = \frac{v^2}{R} \tag{7.12}$$

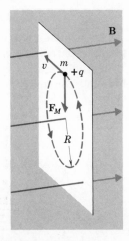

Fig. 7.15 *A charged particle moves in a circular orbit in a uniform magnetic field if* $\mathbf{v} \perp \mathbf{B}$.

Equating these two expressions for a_c and solving for R, we have

$$R = \frac{mv}{q\mathbf{B}} \tag{7.13}$$

Therefore, a charged particle moving at right angles with respect to a uniform magnetic field executes a *circular* orbit in the field with a radius that is directly proportional to its momentum (mv) and inversely proportional to the field strength.

It is important to realize that a charged particle executing an orbit in a static magnetic field in vacuum will neither gain nor lose energy.[1]

Example **7.1**

What is the radius of the orbit of a 1-MeV proton in a magnetic field $\mathbf{B} = 1$ T?

We have

$m = 1.67 \times 10^{-27}$ kg
$q = e = 1.6 \times 10^{-19}$ C

Also, from Example 5.5 we know that the velocity of a 1-MeV proton is

$v = 1.38 \times 10^7$ m/sec. Therefore,

$$R = \frac{mv}{eB}$$

$$= \frac{(1.67 \times 10^{-27} \text{ kg}) \times (1.38 \times 10^7 \text{ m/sec})}{(1.6 \times 10^{-19} \text{ C}) \times (1 \text{ T})}$$

$$= 0.14 \text{ m}$$

[1] In the event that the magnetic field changes with time, the energy of the particle will, in general, be altered; see Section 7.7.

7.6
ORBITS OF
CHARGED
PARTICLES IN
MAGNETIC
FIELDS

One type of device that is often used in the acceleration of charged particles (protons, deuterons, α particles, etc.) to high velocities is the *cyclotron*. The basic idea of cyclotron operation is to accelerate charged particles by means of electric fields while confining the particles with a magnetic field.

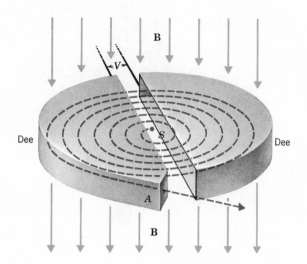

Fig. 7.16 *Schematic of a cyclotron.*

A schematic representation of a cyclotron is shown in Fig. 7.16. The essential elements of a cyclotron are a hollow, cylindrical cavity which is split along a diameter to form two "dees" (so-called because of their shape), an electromagnet (not shown in the figure) which produces a magnetic field perpendicular to the plane of the "dee" structure, and high voltage apparatus which produces a potential difference V between the "dees." Neutral gas atoms (for example, hydrogen) are ionized at the source S, located near the center of the "dees," to produce charged particles that are injected into the left-hand "dee." These particles move in a circular orbit under the influence of the field **B** until they emerge from the left-hand "dee." The particles are then accelerated across the "dee" gap by the electric field and are increased in energy by an amount qV. The electric field exists only *between* the "dees;" the interiors of the conducting "dees" have no electric field. Therefore, when the particles enter the right-hand "dee," they again move in a circular orbit but now of increased radius corresponding to their greater velocity. By the time the particles reach the "dee" gap again, the high-voltage apparatus has switched the polarity of the voltage on the "dees" so that the particles experience another accelerating voltage at the gap. This process is continued for many passages through the gap; each passage increases the energy by the amount qV and the particles spiral outward to greater and greater radii. Near the outer wall of the "dee" structure the particles pass into an *extractor* (usually a pair of plates across which is placed a high voltage) and emerge as a beam of high energy particles at A.

It is essential for the operation of a cyclotron that the voltage across the "dee" gap always be of the correct sign to accelerate the particles rather than to retard them. This is relatively easy to accomplish because of the

ELECTRIC
CHARGES
IN MOTION

important fact that a charged particle moving in a given uniform magnetic field requires a *fixed* time to execute an orbit *independent of its velocity*.[2] Therefore, the particles require the same time to complete each half revolution in the "dees" and arrive at the gap *in phase* with those particles executing orbits with different radii, ready to accept the next accelerating voltage. For a given type of particle and for a given magnetic field there is a single frequency (called the *cyclotron frequency*) at which the polarity of the voltage must be alternated to provide continuing acceleration.

The first cyclotron (only 11 inches in diameter) was constructed by E. O. Lawrence and M. S. Livingston at the University of California in 1930. (Lawrence received the 1939 Nobel Prize in physics for this work.) One of Lawrence's early cyclotrons is shown in Fig. 7.17. Modifications of the basic cyclotron principle have been made to permit the acceleration of particles to ultra-high (relativistic) energies. These machines are known as *synchrocyclotrons* and *synchrotrons*.

Fig. 7.17 *One of E. O. Lawrence's early cyclotrons constructed at the University of California in the 1930s.*

7.7 *Charged Particles in the Earth's Magnetic Field*

THE EARTH'S RADIATION BELTS

During the flight of the satellite Explorer I in 1958, evidence was obtained by James A. Van Allen and his co-workers that there are regions in space near the Earth that contain large numbers of charged particles. These regions are known as the *Van Allen radiation belts* and are regions in which charged particles are trapped in the Earth's magnetic field. Figure 7.18 shows an early map of the radiation belts constructed from satellite data. The various curves represent the surfaces on which the electron density is the same; the numbers are the counts per second that were registered by Van Allen's geiger counters. There are two regions (indicated by the shaded areas) in which the electron density is quite high and these are known as the *inner* and *outer* radiation belts.

---

[2]This statement is only true for velocities sufficiently low that relativistic effects can be ignored.

7.7
CHARGED
PARTICLES IN
THE EARTH'S
MAGNETIC
FIELD

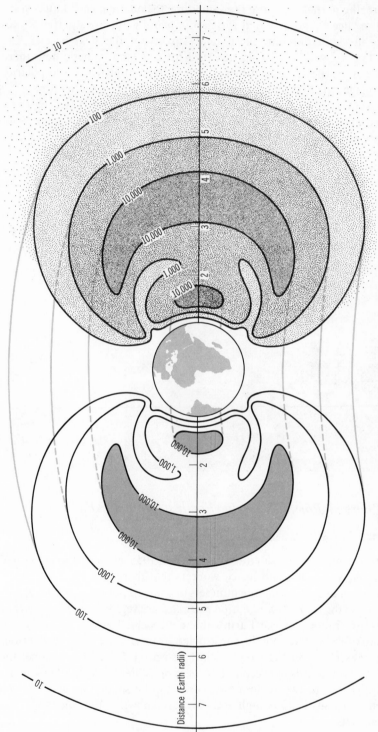

Distance (Earth radii)

Fig. 7.18 *Map of the Van Allen radiation belts constructed from satellite data. The two shaded areas are the inner and outer radiation belts in which the electron density is high.*

A great deal of information regarding the radiation belts has been obtained in recent years from satellite and rocket experiments but, in fact, we do not yet have any really satisfactory theory that describes in detail the data regarding charged particles in the Earth's magnetic field.

AURORA

Another interesting phenomenon that is associated with charged particles in the Earth's magnetic field is the occurrence of the *northern lights* or *aurora*. Auroral displays (see Fig. 7.19) occur most frequently at high northern

Fig. 7.19 *Photograph of an aurora taken in northern Canada from a NASA aircraft flying above the cloud cover. The V-shaped spot on the lower portion of the photograph is Jupiter—the strange shape is caused by the motion of the camera during the 20-sec time exposure.*

latitudes but occasionally they can be observed as far south as the middle of the United States. Auroral light is generated by solar protons that penetrate the Earth's field down to altitudes of about 100 km. At these altitudes the atmosphere is quite tenuous but there are present sufficient numbers of oxygen and nitrogen atoms that collisions between the protons and the atoms produce enough light to be prominently visible. Although auroral activity takes place continuously, the amount of light generated is usually insufficient to be noticeable. On those occasions when there are pronounced solar disturbances in the form of Sun spots and magnetic storms, the number of protons injected into the upper atmosphere is greatly increased and spectacular auroral displays occur.

7.7
CHARGED
PARTICLES IN
THE EARTH'S
MAGNETIC
FIELD

Most stars, including the Sun, have magnetic fields. Because a star consists of rapidly moving and turbulent ionized gases, the magnetic field that is produced by these moving charged particles is not static, as is the Earth's field, but changes with time, frequently in a violent way. When *magnetic storms* occur on a star's surface, great streams of charged particles are usually ejected in solar *flares* or *prominences*. These streams are guided by the star's magnetic field and result in enormous looping flares that rise to heights of thousands of kilometers and then return to the surface. Striking events of this type taking place on the Sun have often been photographed (see Fig. 7.20).

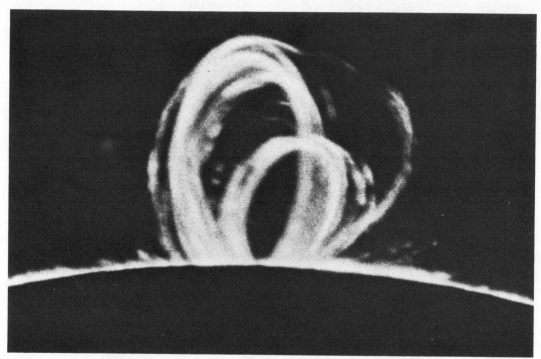

Fig. 7.20 *A solar prominence. Charged particles ejected from the Sun are trapped by the Sun's magnetic field. (Sacramento Peak Observatory, Air Force Cambridge Research Laboratories.)*

7.8 *Fields that Vary with Time*

INDUCED CURRENTS DUE TO THE RELATIVE MOTION OF WIRES AND FIELDS

Thus far we have considered only the effects that are produced when moving charges and currents interact with *static* magnetic fields, that is, fields that do not change with time. We now turn to a discussion of *time-dependent* phenomena, including cases in which there is relative motion between current-carrying wires and magnetic fields or in which a magnetic field varies with time.

ELECTRIC
CHARGES
IN MOTION

156

For the same reason that a moving charge experiences a force in a magnetic field, a current-carrying wire (equivalent to a line of moving charges) that lies in a magnetic field, as in Fig. 7.21, will also be acted on by a magnetic force. If we disconnect the external source (for example a battery) that

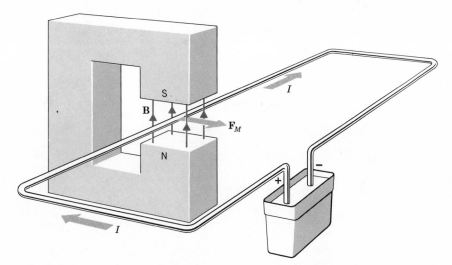

Fig. 7.21 *A static magnetic field exerts a force on a current-carrying wire.*

supplies current to the wire then, of course, the magnetic force will disappear. Now, suppose that we move the current-free wire through the magnetic field in the manner illustrated in Fig. 7.22. This motion will produce a magnetic force on the charges in the wire and they will begin to move along the wire. As long as the wire is in motion in the field, these charges will continue to flow; that is, a current has been *induced* in the wire. The phenomenon of *induction* was discovered in 1831 by the English physicist, Michael Faraday (1791–1867).

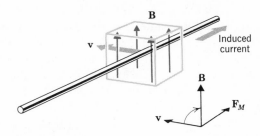

Fig. 7.22 *The motion of a wire through a magnetic field induces a current to flow in the wire. The section of wire shown is only a portion of the complete loop of wire that is necessary in order for a steady current to flow; compare Fig. 7.21.*

In order for a current to be induced in a wire it is not crucial that the field be stationary while the wire moves. It is only the *relative* motion of wire and field that is important. Therefore, in Fig. 7.22, if the wire were stationary and the magnet (and, hence, also the field) moved to the right, a current would be induced to flow in the same direction.

Another situation in which a current is induced by the relative motion

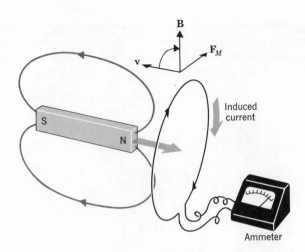

Fig. 7.23 *The relative motion between the field of the bar magnet and the wire loop induces a current in the wire which is detected by the ammeter.*

between a magnetic field and a wire is shown in Fig. 7.23. Here a bar magnet is thrust into a wire loop that is connected to a current-measuring device (an *ammeter*). Notice that the velocity vector, which describes the motion of the wire (and the charges it carries), has a direction opposite to that of the movement of the magnet because this vector describes the velocity of the *wire* relative to the *field*. Application of the right-hand rule for $\mathbf{F}_M$ (for example, at the top of the wire loop, as indicated in Fig. 7.23) shows that a current will flow in a clockwise sense when viewed with the N pole of the magnet approaching. If the magnet continues its motion, the induced current will drop to zero when the magnet is centered in the loop because at that position the motion of the wire (and the charges it carries) is directly *along* the field line; when **v** is parallel to **B**, the magnetic force vanishes. As the S pole of the magnet passes through the loop, an induced current will flow again, but now the direction of flow will be opposite to that shown in Fig. 7.23. (Why?)

CURRENTS INDUCED BY TIME-VARYING FIELDS

Consider a loop of wire that is connected to a battery through a switch. If the switch is open, no current flows and there is no magnetic field. Closing the switch causes the current to flow. But the current does not instantaneously attain its final steady value. Instead, the current is zero at the exact instant that the switch is closed and builds up to its final value during a certain short interval of time. Similarly, the magnetic field that is due to the flow of current starts at zero when the switch is closed and builds up to its final value just as does the current. Therefore, at any particular position in the vicinity of the wire, the magnetic field increases with time during the interval required for the current to attain its final steady value.

We can describe this situation in a pictorial way by saying that the circular magnetic field lines originate at the wire (beginning at the instant that the switch is closed) and spread out into space until the final steady field configuration is attained. The "movement" of the field lines in this outward expansion is similar in its effect to the physical movement of a magnet. Therefore, a current will be induced in a wire that lies in the path of the

ELECTRIC
CHARGES
IN MOTION

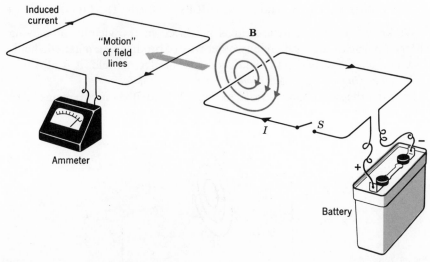

Fig. 7.24 *When the switch S is closed, a current begins to flow in the right-hand loop of wire. A magnetic field expands from this wire as the current builds up; the "moving" field lines induce a current in the left-hand loop.*

"moving" field lines. Figure 7.24, shows such a situation; the field that expands from the right-hand loop when the switch is closed induces a current to flow in the left-hand loop. As soon as the current reaches its final steady value in the right-hand loop, the field ceases to expand and the induced current drops to zero. If the switch is now opened, the magnetic field will collapse and a current will be induced in the left-hand loop but in a direction opposite to that for the case of the expanding field.

A CHANGING MAGNETIC FIELD PRODUCES AN ELECTRIC FIELD

We can summarize in the following way all of the results discussed thus far in this section. If we place a charged particle at a certain position in space and if we allow the magnetic field at that position to change as a function of time, the charged particle will experience a force. The magnetic field can change with time because of the motion of a permanent magnet or current-carrying wire that gives rise to the field or the change with time can be the result of the changing current in a wire. If, instead of a charged particle, we consider a conductor (such as a wire) that contains free electrons, the force exerted on these charges by the changing field will induce a current.

A charged particle at rest in a static magnetic field will remain at rest; an electric field can exert a force on a *stationary* charge but a magnetic field cannot. This must mean that *a changing magnetic field produces an electric field* and it is the electric field that causes the charge to move. Symbolically, we can express this statement as

$$\frac{\Delta \mathbf{B}}{\Delta t} \propto \mathbf{E} \qquad (7.14)$$

We know that a magnetic field is produced in the vicinity of moving charged particles. But is there any other way of producing a magnetic field? Just as a changing magnetic field produces an electric field, it is also true that *a changing electric field produces a magnetic field.*

Consider the situation illustrated in Fig. 7.25. Initially, the two plates carry

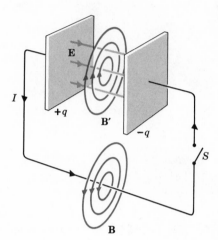

Fig. 7.25 *When the switch S is closed, a current flows in the wire from the positively charged plate to the negatively charged plate. This current produces the magnetic field* **B**; *also, the changing electric field between the plates produces the magnetic field* **B**′. *Notice that the field lines of* **B** *and* **B**′ *have the same direction relative to the direction of current flow.*

charges of $+q$ and $-q$. With the switch in the circuit *open,* the equal and opposite charges on the two plates produce an electric field in the region between them. No current flows in the wire, so there is no magnetic field at any point around the circuit. When the switch is closed, a current begins to flow in the wire and a magnetic field is produced around the wire in the manner that we have previously discussed. But in the region between the plates, there is no flow of ordinary current—there is only a *change* in the electric field because of the buildup of charge on the plates. In spite of the fact that there is no movement of electric charge in this region, nevertheless there is a magnetic field produced as a result of the changing electric field. Since we find experimentally that magnetic fields are produced as the result of current flow as well as changing electric fields, we can summarize the situation symbolically as

$$\boxed{\frac{\Delta \mathbf{E}}{\Delta t} + \text{current} \propto \mathbf{B}} \qquad (7.15)$$

The strengths of the magnetic fields that are due to changing electric fields are, in general, considerably smaller than those due to current flow. Nonetheless, these fields can be detected in favorable situations and the detailed equation relating $\Delta \mathbf{E}/\Delta t$ and **B** can be completely verified.

We have previously stated that magnetic fields have no existence apart from moving electric charges. The fact that magnetic fields can be produced in regions of space where **E** is changing, *even though there are no charged particles to move,* does not invalidate this statement because the electric field

ELECTRIC
CHARGES
IN MOTION

cannot change unless there is a flow of current in another part of the circuit. Therefore, the origin of the magnetic field lies ultimately in the motion of charge.

MAXWELL'S EQUATIONS

Just over a hundred years ago, in 1865, the great Scottish theoretical physicist James Clerk Maxwell (1831–1879) showed that it was possible to base a complete description of electromagnetic field phenomena on a set of only four equations. Much as Newton had drawn on the previous work of others in devising his famous equations of dynamics. Maxwell leaned heavily on the formulations of electric and magnetic phenomena that had been made by others, especially Michael Faraday, in a long series of experimental and theoretical investigations. Thus, Maxwell did not *invent* the equations that now bear his name (in fact, he was responsible for setting down for the first time only *one* of these equations); his important contribution to the subject was to show in a definitive way that these equations form the basis of interpreting all manner of electromagnetic effects (including electromagnetic *waves*, which we shall treat in the next chapter).

Maxwell's four equations can be summarized in words in the following way:

1. Coulomb's law: the electric field of a point charge.
2. Magnetic field lines are continuous and have neither beginning nor end; there are no magnetic monopoles.
3. A changing magnetic field produces an electric field (electromagnetic induction).
4. A magnetic field can be produced by current flow as well as by a changing electric field.

Summary of Important Ideas

The net motion of electric charge constitutes a *current*.

Like magnetic poles *repel;* unlike magnetic poles *attract*.

A freely-suspended magnet aligns itself from S pole to N pole *along* the magnetic field lines.

The magnetic field lines in the space surrounding a magnet have the direction from the N pole to the S pole.

The S pole of the Earth is near the *north* geographic pole.

The poles of a magnet have no independent existence; N and S poles *always* occur together.

The direction of the magnetic field lines surrounding a current-carrying wire is determined by the *right-hand rule*.

Magnetic field lines are *continuous;* they have no beginning and no end. This is equivalent to the statement that *magnetic monopoles do not exist*.

Magnetic fields have no existence independent of moving electric charges.

Elementary atomic magnets are produced by the fields of *circulating electrons*.

The magnetic force $\mathbf{F}_M$ on a moving positively-charged particle is *perpendicular* to the plane defined by $\mathbf{v}$ and $\mathbf{B}$. The direction of $\mathbf{F}_M$ is the same as the direction of advance of a right-hand screw when turned in the sense that rotates $\mathbf{v}$ toward $\mathbf{B}$.

The magnetic force on a charged particle *does no work* (unless the magnetic field changes with time).

A changing magnetic field produces an *electric* field; a changing electric field produces a *magnetic* field.

Maxwell's four equations are a complete description of the electromagnetic field.

Questions

7.1

Explain carefully how a *steady* current can flow in a wire when the electron drift velocity is only 1 mm/sec and the thermal speeds of the electrons are of the order of 10^6 m/sec.

7.2

If you were given two iron bars of identical appearance, one of which is magnetized and one of which is not, how would you decide which is magnetized without using any additional bars or magnets? Would it be possible to make the determination without taking advantage of the Earth's magnetic field?

7.3

An electron is projected into a current loop exactly along the axis. Describe the motion of the electron. What difference will there be if the electron's velocity vector is at a slight angle with respect to the axis of the loop?

7.4

Two identical carboard tubes are wound with wire in exactly the same way. The tubes are placed end-to-end and equal currents are passed through the wires. The currents circulate about the tubes in the same way. Will there be attraction or repulsion between the tubes?

7.5

A current-carrying wire lies in a north-south direction. A compass is placed immediately above the wire and the N pole points eastward. In what direction are the electrons in the wire moving?

7.6

An electron moves in an eastward direction near the equator. In what direction does the Earth's magnetic field exert a force on the electron?

7.7

A wire lies in a north-south direction and a current flows north in the wire. A positively-charged particle moves in the vicinity of the wire. In what direction will $\mathbf{F}_M$ act if (a) the particle is over the wire and moves north, (b) the particle is east of the wire and moves toward the wire, and (c) the particle is west of the wire and moves away from the wire?

7.8

One end of a bar magnet is thrust into a wire loop. The induced current in the wire flows in the clockwise direction as viewed by looking along the direction of motion in the magnet. Which pole of the magnet was thrust into the loop? In what direction will the current flow if the magnet is *withdrawn* from the loop?

7.9

A *short length* of wire is originally at rest in a static magnetic field. A mechanical force is suddenly applied to the wire, causing it to move through the field at a constant velocity. Explain why current will flow in the wire for a short time but will then stop even though the wire continues to move.

Problems

7.1

How much current flows in a 150-W light bulb when connected to a 110-V household circuit? What is the resistance of the light bulb?

7.2

A certain lighting fixture contains five 60-W light bulbs and three 100-Ω light bulbs. How much current does the fixture draw from a household circuit? (The voltage across each bulb is 110 V.)

7.3

Sketch the magnetic field lines for the two pairs of bar magnets shown in the diagram.

7.4

Two long wires lie parallel and carry equal currents in opposite directions. Sketch the lines of **B** in a plane that is perpendicular to the wires. Will the wires be mutually repelled or attracted?

7.5

What is the maximum force that a 1-T magnetic field can exert on an electron whose energy is 10 keV? What is the minimum force and under what conditions would it be attained?

7.6

There are sharp boundaries between a field-free region of space and a region containing a uniform magnetic field with the dimensions shown in the diagram. A charged particle enters the field region (from the field-free region) and moves perpendicular to the field lines. Describe the subsequent motion of the particle for the cases in which the orbit radius, R, has the values: (a) $R < \frac{1}{2}l$, (b) $\frac{1}{2}l < R < l$, and (c) $R > L$.

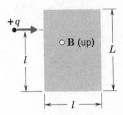

7.7

An electron is accelerated, starting from rest, by falling through a potential difference of 1000 volts. The electron then enters a magnetic field and is found to execute a circular orbit with a radius of 0.2 m. What is the strength of the magnetic field?

7.8

Two particles have the same momentum but one particle carries twice the charge of the other. What will be the ratio of their orbit radii in the same uniform magnetic field?

7.9

A singly-charged carbon ion (C^{12+}) is found to have the same orbit as a 2-MeV proton in a certain magnetic field. What is the energy of the carbon ion in MeV?

7.10

A proton and a deuteron fall through the same potential difference and enter a uniform magnetic field. What is the ratio of the radii of the orbits?

7.11

The *magnetic rigidity* of a charged particle is BR where R is the radius of the orbit that the particle would follow in a magnetic field of strength B. Compare the magnetic rigidity of a 5-MeV proton and a 5-MeV α particle (He^{4++}).

7.12

A 1-g mass is moving eastward at the equator (where $B = 3 \times 10^{-5}$ T) with a velocity $v = 3 \times 10^7$ m/sec. What charge must the mass have if the upward magnetic force cancels the downward gravitational force?

ELECTRIC
CHARGES
IN MOTION

7.13

A cosmic ray proton with an energy of 10^{18} eV behaves as if its mass were approximately 10^9 times its mass when at rest (because of the relativistic increase of mass with velocity). The velocity of such a proton is essentially the velocity of light. Calculate the radius of the orbit that a 10^{18}-eV proton would execute in galactic space where the average magnetic field is about 3×10^{-10} T. Compare the result with the size of the local Galaxy.

7.14

A particle of mass m and charge q moves in a circular orbit in a magnetic field of strength B. Show that the time required to complete an orbit does not depend on the velocity of the particle.

7.15

It is desired that 40-MeV α particles in a cyclotron should execute an orbit with a radius of 1 m. What magnetic field strength is required?

7.16

In a certain region of space there are uniform electric and magnetic fields. The $\mathbf{E}$ and $\mathbf{B}$ field vectors are at right angles with respect to one another. A charged particle with a velocity $\mathbf{v}$ enters this field in a direction perpendicular to both $\mathbf{E}$ and $\mathbf{B}$. Show that the particle will proceed through the field region *undeflected* if $v = E/B$. (Such a device is called a *crossed-field analyzer* and is useful for selecting particles with a given velocity.) If $E = 10^6$ V/m, what value of B is required to allow 1-MeV protons to pass undeflected?

CHAPTER 8
OSCILLATIONS, WAVES, AND RADIATION

Oscillatory and wave phenomena play extremely important roles in all areas of science as well as in our everyday lives. Many of the mechanical devices that we use have oscillating parts—the balance wheel in a clock, the pistons in an automobile engine, and the cutting blade on an electrically-operated jigsaw. Wave motion is also an oscillatory phenomena. As a water wave travels across the sea, water molecules vibrate up and down; as a sound wave travels through the air, air molecules vibrate back and forth. In the case of electromagnetic waves (radio waves, light, X rays, etc.), the electromagnetic field oscillates as the wave moves forward. As we progress to the later chapters we shall see the great importance of wave phenomena in modern and contemporary physics. We shall even see that *material particles* can also be described in terms of *waves,* a fact that is of fundamental importance in the quantum theory of matter.

8.1 *Simple Harmonic Motion*

HOOKE'S LAW

The oscillatory motion of a mass attached to a coiled spring is the prototype of large and important class of oscillatory phenomena called *simple harmonic motion*. Figure 8.1a shows such a mass at rest in its equilibrium position on a frictionless surface. If we apply an external force to displace the mass to the right, and then remove the external force, there will be a restoring force $\mathbf{F}$ exerted on the mass by the spring and directed to the left (Fig. 8.1b). If the displacement is not too great, the elastic restoring force is directly proportional to the magnitude of the displacement; that is,

$$F = -kx \tag{8.1}$$

where the negative sign means that the direction of the force is opposite to the direction of displacement. If the initial displacement were to the *left,* the spring would be compressed and the restoring force would be to the *right*.

Many elastic materials, if not stretched too far, obey the simple linear

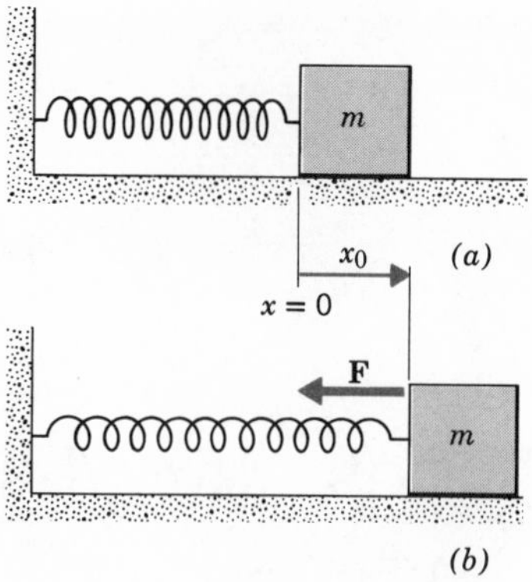

Fig. 8.1 (a) A mass m rests on a frictionless surface and is attached to a spring: the mass and spring combination is in its normal (equilibrium) condition. (b) If m is displaced to the right an amount x_0 (by an external force), there will be a restoring force to the left given by $F = -kx_0$, where k is the force constant characteristic of the particular spring.

relationship between force and displacement given in Eq. 8.1. This relationship is called *Hooke's law,* after Robert Hooke (1635–1703), an English scientist and a contemporary of Newton.

A GRAPHICAL DESCRIPTION OF THE MOTION

If the mass is released from its position of initial extension ($x = x_0$), it will be accelerated to the left by the restoring force. At $x = x_0$ the acceleration is a maximum; as the mass moves toward $x = 0$ the velocity increases while the acceleration decreases. When the mass reaches $x = 0$, the restoring force (and, hence, the acceleration) will have decreased to zero, but the velocity of the mass will have been increased to its maximum value at this point and the inertia of the mass will carry it into the region of negative x (to the *left* of $x = 0$). In this region the restoring force (and, hence, the acceleration) is directed to the *right* and the mass will be slowed down. At $x = -x_0$, the motion will stop and the acceleration (which is still toward the right) will cause the mass to reverse its motion and move toward $x = x_0$ again. The entire process is one of *cyclic* (or *oscillatory* or *periodic*) motion, with the mass vibrating back and forth between $x = x_0$ and $x = -x_0$.

We can obtain a record of the motion of the mass as a function of time in the following simple way. As shown in Fig. 8.2, we attach a pen to the mass and allow it to touch a roll of paper that is moved uniformly in a direction perpendicular to the direction of motion of the mass. In this way we obtain a displacement-time graph of the motion. Examination of the graph shows that it is the same as a *sine* or *cosine* curve, familiar in trigonometry. After every interval of time τ, which is called the *period,* the motion repeats itself (see Fig. 8.3). The quantity x_0 is the *amplitude* of the motion, the maximum excursion from the equilibrium position experienced by the mass. A sinusoidal function (sine or cosine) varies in a simple and regular way (that is, the variation is *harmonic*), and therefore motion described by such functions is termed *simple harmonic motion.*

Figure 8.3 shows an oscillatory motion that continues indefinitely without any change in amplitude. This is the ideal case in which frictional losses are ignored. Of course, in a real situation, friction will be present and the motion will not persist forever unless energy is continually supplied to the system. Friction causes the amplitude of the motion to decrease with time (this is called *damping*), and eventually the oscillations will cease.

OSCILLATIONS, WAVES, AND RADIATION

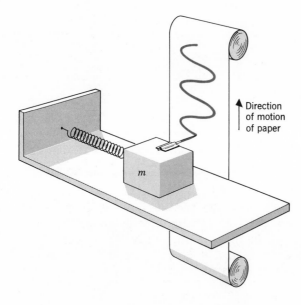

Fig. 8.2 *A simple method for recording the motion of an oscillating mass as a function of time. In a real case, friction would be present and the motion would die away (or* damp out*).*

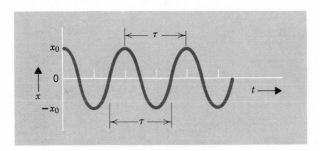

Fig. 8.3 *The displacement as a function of time for a mass undergoing simple harmonic motion. There is a time interval τ (the period) between any two successive corresponding points on the curve.*

THE PERIOD AND FREQUENCY

If we increase the mass that is attached to the spring, the motion becomes more sluggish and consequently the period *increases;* if the mass is increased by a factor of four, we find that the period increases by a factor of two. That is, $\tau \propto \sqrt{m}$. Similarly, if the force constant of the spring is increased, the motion is speeded up and the period *decreases.* In this case we find $\tau \propto 1/\sqrt{k}$. The general expression for the period of a simple harmonic oscillator is

$$\tau = 2\pi \sqrt{\frac{m}{k}} \qquad (8.2)$$

The *period* of a vibratory motion is the time interval required for one complete oscillation of the system. The *frequency* of a vibratory motion is the number of complete oscillations that the system makes in unit time. It is customary to use the symbol ν to denote frequency. The unit of frequency is *cycles/sec* (or simply *sec*$^{-1}$), the abbreviation for which is *Hertz (Hz)*, in honor of Heinrich Hertz (1857–1894) who made great contributions to the study of electrical oscillations. If the period of a system is τ, it is clear that the system will experience $1/\tau$ vibrations per second. That is, the period and the frequency are reciprocally related:

$$\nu = \frac{1}{\tau} \qquad (8.3)$$

In Eq. 8.2 notice that the initial displacement x_0 of the oscillator does not appear. That is, the period is *independent* of the amplitude of vibrations. This fact is characteristic of all types of simple harmonic oscillators.

Example **8.1**

Suppose that a mass of 0.8 kg is attached to a spring that requires a force of 2 N to extend it to a length 0.5 m greater than its natural length. What is the period of the simple harmonic motion of such a system?

The force constant is

$$k = \frac{F}{x} = \frac{2\,\text{N}}{0.5\,\text{m}} = 4\,\text{N/m}$$

Therefore, the period is:

$$\tau = 2\pi \sqrt{\frac{m}{k}} = 2\pi \sqrt{\frac{0.8\,\text{kg}}{4\,\text{N/m}}} = 2\pi \sqrt{0.2\,\text{sec}^2}$$

$$= 2\pi \times (0.45\,\text{sec}) = 2.8\,\text{sec}$$

and the frequency is

$$\nu = \frac{1}{\tau} = \frac{1}{2.8\,\text{sec}} = 0.35\,\text{Hz}$$

8.2 *Wave Motion*

THE PROPAGATION OF WAVE PULSES

The oscillatory motions we have just discussed involve the motions of only a single particle or mass. We now turn our attention to situations in which the motion of any given particle influences and is influenced by the motion of its neighbors. An important case of such a cooperative phenomenon is that of *wave motion*.

Examples of wave motion are to be found virtually everywhere. Water waves travel across the seas. Sound is propagated by waves in the air. A piano or a violin produces its characteristic sounds by the wave motions of strings. The air (and even empty space) is permeated by electromagnetic waves in the form of radio waves and light. In all cases of mechanical waves (as distinct from electromagnetic waves, which are due to the periodic variations of the electromagnetic field), the vibratory motion of particles is involved. But even though these waves propagate through the air or along a string or across the ocean, the individual motions of the particles are never very large. For example, Fig. 8.4 shows a wave disturbance propagating along a coiled spring. A ribbon, tied to one of the coils, provides a marker

OSCILLATIONS, WAVES, AND RADIATION

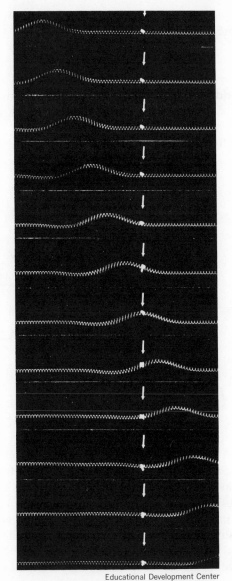

Fig. 8.4 *A wave pulse travels along a coiled spring from left to right. The various pictures are individual frames from a film of the motion recorded by a movie camera. A ribbon is tied to one of the coils* (indicated by the arrow) *in order to illustrate the motion of a particular small portion of the spring.*

for observing the motion of a particular portion of the spring. It is evident that no portion of the spring moves very far in the horizontal direction and yet the pulse travels along the spring by virtue of the fact that the particles at the front of the disturbance are forced to move upward and those at the rear are forced to move downward to their original positions.

TRAVELING WAVES

If, instead of setting up a single pulse on a spring or string, we move one end up and down in a regular (sinusoidal) fashion, then a regular series of pulses will be propagated along the string; that is, a *traveling wave* will be generated. The velocity of the wave propagation can be found by following one wave crest as a function of time. If the peak moves forward a distance Δx in a time Δt, the propagation velocity is

$$v = \frac{\Delta x}{\Delta t} \tag{8.4}$$

The period τ of the driving oscillation is also the period of the wave motion. During a time interval τ the wave moves forward by an amount called the *wavelength* λ of the wave (Fig. 8.5). Therefore, if $\Delta t = \tau$, then $\Delta x = \lambda$, and the important characteristics of the wave are related according to

$$\boxed{v = \frac{\lambda}{\tau} = \lambda \nu} \tag{8.5}$$

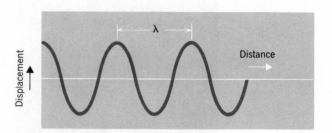

Fig. 8.5 *As a traveling wave moves forward, successive crests are separated by one* wavelength *of the wave.*

SUPERPOSITION

Wave motion is another physical phenomenon that obeys the principle of superposition (see Sections 6.2–6.4). If two separate pulses travel toward one another along the same spring, their motions are completely independent. Figure 8.6 is a photographic record of two pulses on a coiled spring that approach and pass one another. After passing, they continue in opposite directions without having suffered any change in shape or size (or velocity). When the two pulses pass one another, the displacement of the spring is the *sum* of the two individual displacements. If these individual displacements are of the same shape and size but are of opposite signs, then a *cancellation* will result at the moment of passing (see the 5th frame of Fig. 8.6). If the pulses are of the same sign, then they will *add* at the moment of passing.

TRANSVERSE AND LONGITUDINAL WAVES

In the wave motions we have been discussing, the particles in the strings or springs move at right angles to the direction of propagation of the wave. Such waves are therefore called *trasverse* waves. Wave motion is also possible in which the particles move back and forth along the direction of wave propagation. Such waves are called *longitudinal* waves. An example of this type of wave motion is the *compressional* waves that can be propagated in a spring (see Fig. 8.7a). At any instant, portions of the spring are alternately compressed and extended, and there is a regular variation along the length of the spring. A similar situation exists in the *sound* waves that can be propagated in a column of gas (see Fig. 8.7b). In this case the *density* of the gas molecules (and the gas pressure) varies regularly along the column.

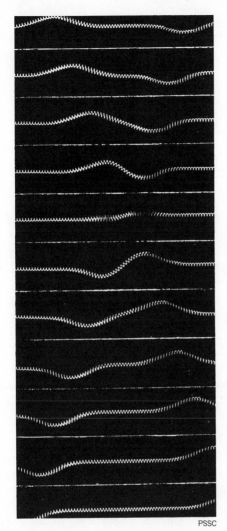

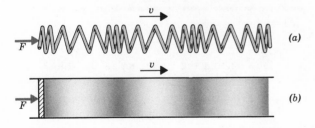

Fig. 8.6 *The superposition of two almost identical wave pulses traveling along a coiled spring. At the moment of passing (fifth frame), they almost cancel each other. Notice, however, that the blurring of the photograph indicates that, although the spring has almost no net displacement, there is a substantial vertical velocity at two positions.*

PSSC

(a)

(b)

Fig. 8.7 *(a) Longitudinal compressional waves in a coiled spring are initiated by the application of a periodic driving force at one end. (b) Longitudinal waves (sound waves) in a column of gas are initiated by the application of a periodic force to a piston located in one end.*

SOUND WAVES

The propagation of sound waves through the air (at standard conditions) takes place with a velocity of approximately 330 m/sec or 1100 ft/sec, which, within wide limits, is independent of the frequency of the wave. When such a pressure wave reaches our ears, it produces vibrations in the ear's membranes which provoke a nervous response and we *hear* the sound. But a hearing sensation is produced in the human nervous system only if the

8.2
WAVE
MOTION

frequency of the wave is between ~16 Hz and ~20,000 Hz. (The upper limit tends to decrease with age.) For these extreme frequencies, the corresponding wavelengths are

$$\lambda_1 = \frac{v}{\nu_1} = \frac{330 \text{ m/sec}}{16 \text{ sec}^{-1}} \cong 20 \text{ m}$$

$$\lambda_2 = \frac{v}{\nu_2} = \frac{330 \text{ m/sec}}{2 \times 10^4 \text{ sec}^{-1}} \cong 0.016 \text{ m} = 1.6 \text{ cm}$$

The velocity of propagation of sound waves is different in different media. Some representative values are given in Table 8.1.

Table **8.1** *Sound Velocities in Some Materials*

Substance	Velocity (m/sec)
Granite	~6 × 10³
Iron	5.13 × 10³
Sea water	1.53 × 10³
Lead	1.23 × 10³
Air	3.31 × 10²

Example **8.2**

What is the frequency of a 2-cm sound wave in sea water?

From Table 8.1, $v = 1.53 \times 10^3$ m/sec in sea water, so

$$\nu = \frac{v}{\lambda} = \frac{1.53 \times 10^3 \text{ m/sec}}{0.02 \text{ m}}$$

$$= 7.6 \times 10^4 \text{ Hz} = 76 \text{ kHz}$$

which is an *ultrasonic* wave (that is, above the human audible range). A 2-cm sound wave in *air* would be audible. (What is the frequency?)

THE DOPPLER EFFECT

We are all familiar with the fact that the frequency of the sound from a siren on a moving vehicle changes dramatically when the vehicle passes. When the sound source is approaching, the apparent frequency is high, and, when the source passes and is moving away, the apparent frequency is low. The dependence of the frequency of a wave disturbance on the relative motion of the source and the observer is termed the *Doppler effect,* after the Austrian physicist, Christian Johann Doppler (1803–1853), who extensively studied this phenomenon.

Figure 8.8 shows the pattern of circular waves spreading out from a vibrating rod that is moving through water. The waves are bunched up to the right because the source of the waves (the vibrating rod) is moving in this direction. An observer on this side would measure a shorter wavelength (that is, a higher frequency) for the waves than if the source were at rest.

OSCILLATIONS, WAVES, AND RADIATION

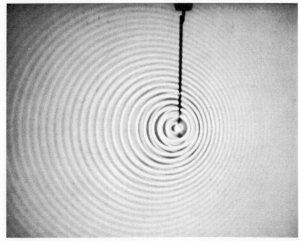

Fig. 8.8 *Photograph of water waves produced by a moving source. The dark line in the photograph is a vibrating wire that is moved through the water from left to right.*

Similarly, the waves to the left of the source are spread out; the wavelength is longer and the frequency is lower.

The Doppler effect exists for all types of wave motion—sound waves in air, elastic waves on springs or in solids, water waves, etc., even electromagnetic waves (for example, *light*). Indeed, we have already seen (Section 1.4) that measurements of the Doppler shifts in the light from galaxies constitute one of the most important methods for determining the distances to the remote regions of the Universe.

STANDING WAVES

Wave propagation on a long, unterminated string consists of traveling waves, a regular sequence of sine-like wave patterns that move uniformly along the string. If we now consider the wave motion of a string that is attached to a rigid support at each end, we find a distinctly different type of wave motion. Since the ends are held in fixed positions, it is clear that the wave pattern must be such that the ends always correspond to points of zero displacement (that is, the ends are *nodes*). The simplest wave pattern for a terminated string is one-half of a sine curve with zero displacement at the ends and a maximum displacement in the center (Fig. 8.9*a*). The wavelength of such a *standing wave* is exactly twice the distance between the supports: $\lambda = 2L$. Waves with shorter wavelengths are also possible, but in order for the ends to be nodes, there must always be an integer number of half wavelengths between the supports (Fig. 8.9). Thus, the condition for standing waves of wavelength λ on a string of length L is

$$n \frac{\lambda}{2} = L, \qquad n = 1, 2, 3, \ldots \qquad (8.6)$$

Standing waves occur for situations other than transverse vibrational waves on stretched strings. Standing sound waves can be set up between reflecting walls, and electromagnetic waves (including light) can be reflected from a pair of surfaces to form standing waves.

8.2
WAVE
MOTION

175

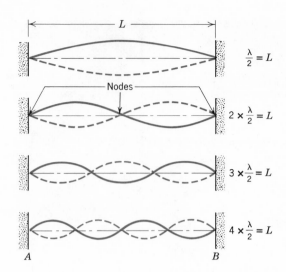

Fig. 8.9 *Standing waves on a stretched string. All such waves must have nodes at the termination points.*

8.3 *Diffraction*

PLANE WAVES

The waves we have been discussing move along straight lines guided by strings or springs; such waves are called *one-dimensional waves*. There are also important types of wave motions that take place in two or three dimensions. Suppose we place a long, straight board in still water, with its longest dimension horizontal, so that part of the board is submerged. now we oscillate the board back and forth with a regular periodic motion that is transverse to the board's longest dimension (Fig. 8.10). Each time the board

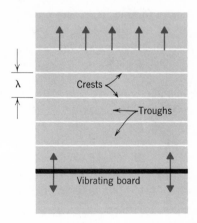

Fig. 8.10 *A vibrating board in an expanse of still water produces two-dimensional plane waves.*

moves, it piles up the water in front of it and pushes the water forward. As a result, a series of crests are propagated across the water with troughs between them. This type of movement constitutes a wave motion and the distance between successive crests (or troughs) is the wavelength of the disturbance. Waves that propagate in this manner are called two-dimensional *plane waves*.

OSCILLATIONS, WAVES, AND RADIATION

Suppose that a plane wave of wavelength λ is incident on a panel into which is cut a slot of width d, where d is small compared to λ (see Fig. 8.11*a*). Will the wave that emerges from the slot be confined to the narrow region of width d? The answer to this question is "no," and the reason is that waves exhibit the phenomenon of *diffraction*. A convenient way to view this effect is to use a clever construction invented by the Dutch mathematical physicist, Christiaan Huygens (1629–1695), who first formulated the wave theory of light. According to *Huygens' principle,* every point on a wavefront can be considered as a source of circular waves (or spherical waves in the case of three dimensions). In the event that the length of the wavefront is small compared to the wavelength of the wave, as in Fig. 8.11, the slot becomes essentially a point source. Then, the wave pattern to the right of the panel consists simply of circular waves. Figure 8.11*b* shows a photograph confirming this diffraction effect for the case of a narrow slot.

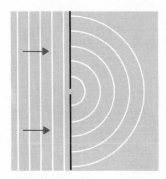 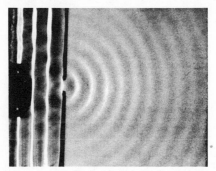

Physics I by Elisha Huggins Benjamin

Fig. 8.11 *Pattern of waves produced by a plane wave incident on a slot. Even though the width of the slot is actually not much smaller than the wavelength of the incident wave, nevertheless, a circular wave pattern results. (The dark rectangle at the left of the photograph is a part of the mechanism that produces the plane wave.)*

If the width of the slot is large compared to the wavelength of the incident disturbance ($d \gg \lambda$), then the Huygens' construction shows that the wave pattern to the right of the panel is essentially a plane wave, with curved ends (Fig. 8.12). Thus, the wave is largely unaffected by the presence of the panel but the circular wave effect persists near the edges of the slot.

8.4 *Interference*

TWO-SOURCE INTERFERENCE

By combining the principle of superposition with Huygens' principle, we are able to explain a variety of important and interesting *interference* effects that occur in wave motions of all sorts. Any wave motion in which the amplitudes of two or more waves combine will exhibit *interference*. Wave pulses of opposite signs that are traveling along a string will *cancel* when they pass (Fig. 8.6); we call this effect *destructive interference*. If the pulses

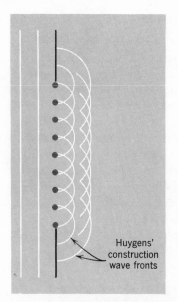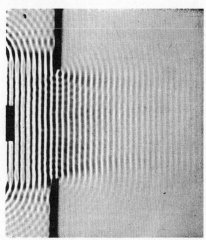

Huygens'
construction
wave fronts

Physics I by Elisha Huggins Benjamin

Fig. 8.12 *If the width of the slot is large compared to the wavelength of the incident plane wave, the wave passes through the slot almost unaffected. There is a noticeable diffraction effect only near the edges of the slot.*

are of the same sign, they will *add* when passing; this is *constructive interference*.

If we allow a plane wave to strike a panel in which there are two slots (each with $d \ll \lambda$), Huygens' principle tells us that these slots will act as separate sources of circular waves. At certain definite positions on the surface of the water, the amplitudes of the waves from the two sources will have the same sign (the waves will be *in phase*) and constructive interference will result; that is, at these positions the disturbance of the water will be enhanced. At other positions the waves will have opposite signs (the waves will be *out of phase*) and destructive interference will result; that is, at these positions the water will remain calm. Figure 8.13 shows the pattern of constructive and destructive interference for such a case.

If, instead of water waves, we allow light of wavelength λ to be incident on a pair of narrow slots in a panel, then on a screen a distance L away we find a series of equally spaced bright and dark lines that result from constructive and destructive interference (Fig. 8.14). The spacing Δx between successive bright or dark lines is directly proportional to λ and to L, and inversely proportional to the distance D that separates the slits; that is,

$$\Delta x = \frac{\lambda L}{D} \qquad (8.7)$$

Because the spacing Δx depends on the wavelength, if we use light consisting of two different colors (say, red and blue), the bright red lines and the bright blue lines will occur at different positions (except for the central line). The wavelength of red light is greater than the wavelength of blue light, so the spacing of the blue lines will be smaller than the spacing

OSCILLATIONS,
WAVES, AND
RADIATION

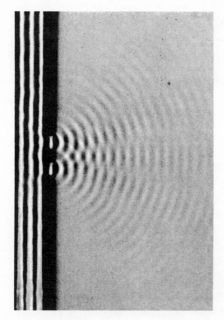

Fig. 8.13 *Photograph of the interference pattern in water waves produced by a plane wave incident on a pair of slots.*

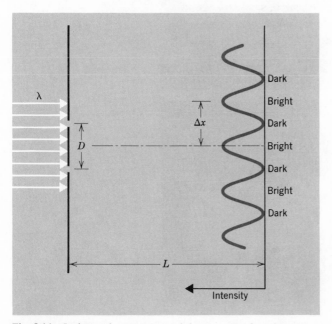

Fig. 8.14 *Light incident on a pair of slots in a panel produces a series of bright and dark lines on a screen a distance L away.*

of the red lines. If *white* light (that is, light consisting of *all* colors) is used, only the central bright line will be prominent—on either side of this line there will be a jumbled blur of colors.

Photographs of the interference patterns obtained for slits with three different separations are shown in Fig. 8.15.

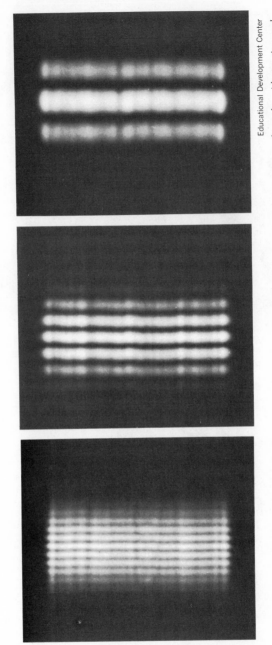

Fig. 8.15 *Double-slit interference patterns for slits with three different separations. Which case corresponds to the wide spacing and which to the narrow spacing?*

Educational Development Center

Example **8.3**

In a double-slit experiment, $D = 0.1$ mm and $L = 1$ m. If yellow light is used, what will be the spacing between adjacent bright lines?

The wavelength of yellow light is approximately 6×10^{-7} m (see Section 8.5). Therefore, the spacing is

$$\Delta x \cong \frac{\lambda L}{D} = \frac{(6 \times 10^{-7} \text{ m}) \times (1 \text{ m})}{10^{-4} \text{ m}} = 0.006 \text{ m} = 6 \text{ mm}$$

Thus, the spacing between lines is about 6 mm or $\frac{1}{4}$ of an inch.

SINGLE-SLIT DIFFRACTION

If we allow a plane wave to illuminate a *single* slit of narrow width, we also find a regular pattern of constructive and destructive interference. The reason for the occurence of interference in this case is that each portion of the slit opening acts as a source of secondary waves. (In the double-slit cases just described, the slit opening was considered to be so narrow that we could assume only one secondary wave originated at each slit.) Then, the interference among these various secondary waves produces the single-slit interference pattern. Figure 8.16 shows a photograph of water waves incident on a slit which has a width only slightly larger than the wavelength of the incident wave. It is apparent that there is a broad central maximum in the pattern of transmitted waves. Closer examination will reveal that there are also weak secondary maxima with interference minima between them.

Light waves exhibit a similar effect. Figure 8.17 shows a photograph of the diffraction pattern on a screen located behind a single slit illuminated with light. This photograph clearly shows that the central maximum is considerably broader than the secondary maxima that lie on either side. In fact, a detailed analysis shows that the spacing between successive dark lines on either side of the central maximum is

$$\Delta x = \frac{\lambda L}{d} \qquad \text{(dark lines)} \tag{8.8}$$

where d is the width of the slit.

The graph above the photograph gives, in a schematic way, the distribution of light intensity along the screen. The intensities shown for the secondary maxima have been enlarged to show the pattern in more detail: these secondary maxima are actually quite weak.

8.5 *Electromagnetic Radiation*

ENERGY TRANSFER BY ACCELERATED CHARGES

From the discussions in the preceding chapter, we know that a steady current flowing in a wire will produce a static (that is, unchanging) magnetic field in the vicinity of the wire. If a charged particle moves in such a field, the force exerted on the particle by the field is always at right angles to the direction of motion of the particle. Therefore, although the field can

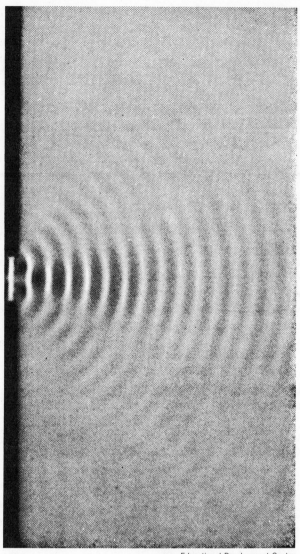

Fig. 8.16 *Interference pattern produced by water waves incident on a single slit. There is a broad central maximum with weaker secondary maxima on either side.*

Educational Development Center

accelerate the particle (by causing it to move in a circular orbit), the *speed* of the particle is not changed. A static magnetic field can do no *work* on the particle and there can be no *energy* transferred from the moving charges in the wire to the particle via the intermediary of the field.

Now let us consider the case in which the current in the wire is allowed to vary so that the magnetic field is no longer static. A changing current produces a changing magnetic field. We know that a changing magnetic field produces an electric field and that this electric field can act on a charged particle to change its energy. Therefore, energy can be transferred from a wire carrying a changing current to a charged particle in the vicinity of the wire, although no such transfer can occur for the case of a wire carrying a steady current. The essential difference in the two situations is that the charges that move in the wire in the case of the changing current undergo accelerations, whereas there is no acceleration of the moving charges in the

OSCILLATIONS, WAVES, AND RADIATION

182

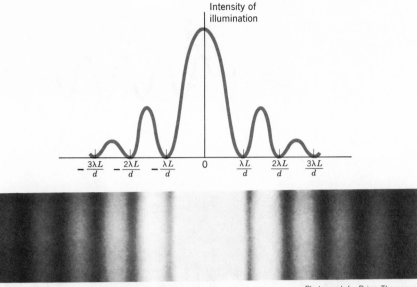

Intensity of illumination

$-\dfrac{3\lambda L}{d}$ $-\dfrac{2\lambda L}{d}$ $-\dfrac{\lambda L}{d}$ 0 $\dfrac{\lambda L}{d}$ $\dfrac{2\lambda L}{d}$ $\dfrac{3\lambda L}{d}$

Photograph by Brian Thompson

Fig. 8.17 *The diffraction pattern, produced by a single slit. The photograph has been over-exposed in order to reveal the secondary maxima; therefore, the central maximum is "washed out" and does not appear in its full intensity. The intensity graph at the top is only schematic, the secondary maxima are actually considerably less intense than shown.*

case of the steady current. *Only accelerating charges can produce energy transfers through the electromagnetic fields that they generate.*

ELECTROMAGNETIC WAVES

It is relatively easy to construct a current source that provides a regularly varying current. Such a current flows first in one direction along the wire and then in the opposite direction. (In fact, ordinary household electrical systems carry such *alternating current* or AC; the standard United States' frequency for AC current is 60 Hz.) Such current variations clearly constitute a case of accelerated motion of the charges in the wire and can therefore lead to the transfer of energy to charged particles outside the wire.

The generation of an electromagnetic field by the changing current in a wire is similar to the production of a traveling wave in a string by applying an oscillation to one end. Just as a regular displacement in the string is propagated along the string, the regular variations of the field vectors, **E** and **B**, are propagated through space (Fig. 8.18).

In a region of space that is far from the source of the radiation (for example, a broadcast antenna), the outward propagating electromagnetic wave is a *plane wave*. Electromagnetic waves have the following important properties (see Fig. 8.19):

1. The electromagnetic field vectors, **E** and **B**, in a *plane wave* are everywhere mutually perpendicular:

$$\mathbf{E} \perp \mathbf{B} \quad \text{(plane wave)} \tag{8.9}$$

2. The direction of propagation of an electromagnetic wave is given by a right-hand rule: the direction of propagation is the same as the direction

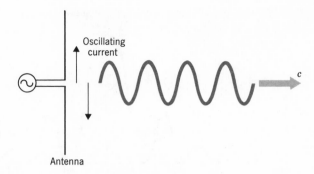

Oscillating
current

Fig. 8.18 *The sinusoidal variation of current flowing in a wire produces sinusoidal variations in* **E** *and* **B** *which are propagated through space with the velocity* **c**. *Here, the source of the electromagnetic wave is a simple antenna.*

Antenna

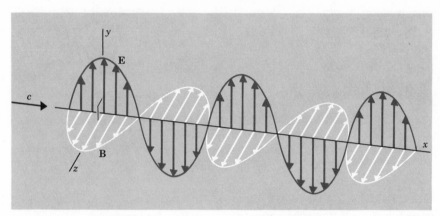

Fig. 8.19 *At a large distance from the antenna (located in the negative* x-*direction), the variation of* **E** *and* **B** *with distance at a given instant is sinusoidal. Furthermore, the electromagnetic wave is a plane wave and has* **E** ⊥ **B** *at every point and for all values of the time.*

of advance of a right-hand screw when turned in the sense that carries **E** into **B** (Fig. 8.20). Even though the directions of the vectors **E** and **B** change with time and with position, the wave always propagates in the same direction. Application of the right-hand rule to the various portions of the wave in Fig. 8.19 shows this to be the case.

3. Electromagnetic waves are *transverse waves*. In Fig. 8.19 notice that the vectors **E** and **B** are always *perpendicular* to the direction of propagation of the wave; at large distances from the source, the field vectors never have any component in the direction of propagation.

4. The velocity of propagation in empty space of all types of electromagnetic waves (light, radio waves, X rays, etc.) is $c = 3 \times 10^8$ m/sec.

5. Electromagnetic waves carry *energy* and *momentum*.

That electromagnetic radiation carries energy should come as no surprise. After all, energy from the Sun carried to the Earth by electromagnetic radiation supplies the light and heat necessary to sustain life on Earth. But it is more difficult to appreciate the fact that electromagnetic waves have momentum as well because the momentum transferred by electromagnetic radiation is always very small, and, in fact, it is quite difficult to measure the effects of electromagnetic momentum in the laboratory. There is, however, a striking effect, readily observable, that is the direct result of the

OSCILLATIONS,
WAVES, AND
RADIATION

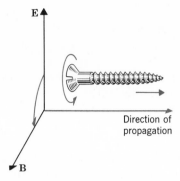

Fig. 8.20 *The direction of propagation of an electromagnetic wave is the same as the direction of advance of a right-hand screw when it is turned in the sense that carried* **E** *into* **B**.

transfer of momentum by electromagnetic radiation. Comets are members of the solar system which are composed of swarms of solid particles, rocky material, and frozen gases.[1] Cometary orbits are highly elongated, and so most of the time a comet is far from the Sun. When it approaches the Sun, however, the solar radiation vaporizes and "boils off" some of the cometary material. This material is subject to the effects of electromagnetic momentum carried by the solar radiation (*radiation pressure* or *light pressure*). As a result, the vaporized material is forced away from the comet and is visible as the cometary *tail*. The effect of radiation pressure is always to force the tail of a comet to point away from the Sun (Fig. 8.21), so that the tail can even

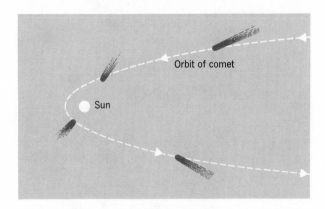

Fig. 8.21 *Because of radiation pressure (transfer of electromagnetic momentum), cometary tails always point away from the Sun.*

precede the head of the comet through space. This behavior of cometary trails have long been known (even in Kepler's time), but the explanation was lacking until it was recognized to be the result of momentum carried by electromagnetic radiation.

THE SPECTRUM OF ELECTROMAGNETIC RADIATION

In 1862, James Clerk Maxwell, predicted on the basis of his theory of electromagnetism that electromagnetic *waves* should exist. His calculations showed that these electromagnetic waves should propagate with a velocity that was the same as that previously found for the propagation of light in

[1]The astronomer Fred Whipple has said that a comet resembles a "dirty iceberg."

air (or empty space). This fact immediately suggested that light is just a particular form of an electromagnetic wave.

Electromagnetic waves (apart from *light*) were not observed until 1887 when Heinrich Hertz produced waves with wavelengths of 10–100 m by forcing a charged sphere to spark to a grounded sphere. Hertz's experiments indicated that these waves were identical in every respect to light waves, except that the wavelength was much longer.

Following the pioneering experiment of Hertz, electromagnetic waves were generated with an ever-increasing range of frequencies. Indeed, it seemed that waves with *any* frequency could be produced if some method of driving electric charges with the appropriate oscillation frequency could be found. This is, in fact, the situation today: there appears to be no physical limitation on the frequency of electromagnetic waves—all we require is a suitable source. Electronic methods have been used to generate electromagnetic waves with frequencies up to about 10^{12} Hz. In this range of frequencies we classify the radiation as either *radiofrequency* (RF) waves or as *microwaves* (see Fig. 8.22). In the former category are the standard broadcast,

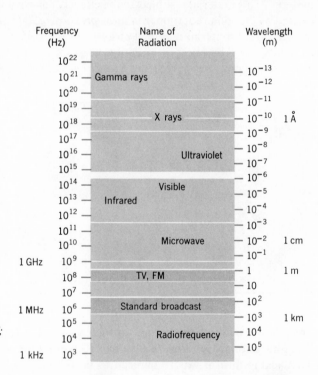

Fig. 8.22 *A part of the spectrum of electromagnetic radiation. The classification terms used for electromagnetic radiations are not well-defined; there is actually considerable overlapping of adjacent categories.*

FM, TV, air and marine, and amateur broadcast bands. Radar and point-to-point relay signaling use microwaves.

In order to generate radiation with frequencies above the microwave range, direct electronic methods are no longer useful, and we employ *atomic* radiations. *Infrared* or *heat radiation* lies in the frequency range between microwaves and the narrow band of frequencies that constitute *visible* radiation. At still higher frequencies are *ultraviolet* radiation and *X* rays. The limit on the frequency that can be generated by atomic systems lies near

OSCILLATIONS,
WAVES, AND
RADIATION

186

Table **8.2** *Some Typical Radiations in and near the Visible Spectrum*

Name	Wavelength (m)	Wavelength (Å)	Frequency (Hz)
Near infrared	1.0×10^{-6}	10,000	3.0×10^{14}
Longest visible red	7.6×10^{-7}	7,600	3.9
Orange	6.1×10^{-7}	6,100	4.9
Yellow	5.9×10^{-7}	5,900	5.1
Green	5.4×10^{-7}	5,400	5.6
Blue	4.6×10^{-7}	4,600	6.5
Shortest visible blue	4.0×10^{-7}	4,000	7.5
Near ultraviolet	3.0×10^{-7}	3,000	10
X ray (long wavelength)	3.0×10^{-8}	30	1,000
X ray (short wavelength)	1.0×10^{-11}	0.1	300,000

10^{20} Hz; radiation with higher frequencies (*gamma rays*) are produced within *nuclei*.

The fact that names have been assigned to these various bands of radiation must not obscure the essential feature of electromagnetic waves, namely, that all of these radiations are *identical* in character and differ only in their *frequency*. Thus, the radio waves that are broadcast from a 500-ft radio tower are exactly the same as the penetrating gamma rays that originate in a nucleus whose diameter is only 10^{-14} m; the only difference between the two radiations is a factor of 10^{15} or so in frequency.

SOLAR RADIATION AND THE SENSITIVITY OF THE EYE

When electromagnetic radiation from the Sun strikes the Earth's atmosphere, a portion is reflected back into space (particularly from cloud formations), and some is absorbed by the molecules in the air. The atmosphere

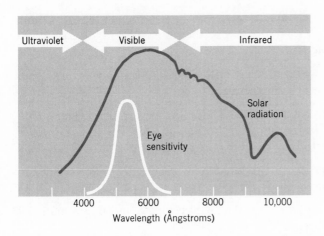

Fig. 8.23 *The spectrum of solar radiation at the surface of the Earth compared with the sensitivity of the human eye. (The solar radiation curve contains numerous dips corresponding to the selective absorption of certain wavelengths by molecules in the air.)*

is an effective shield against the Sun's ultraviolet radiation and very little of this radiation reaches the surface of the Earth. On the other hand, the atmosphere is quite transparent to infrared radiation. Consequently, the spectrum of solar radiation at the Earth's surface (Fig. 8.23) shows a con-

8.5
ELECTRO-
MAGNETIC
RADIATION

centration of energy in the infrared region, with very little in the ultraviolet region.

Figure 8.23 also shows the way in which the sensitivity to radiation of the human eye varies with wavelength. The eye is most sensitive to blue-green light ($\lambda \cong 5500$ Å) and has almost zero response to radiation with wavelengths outside the narrow range from 4000 Å to 7000 Å. The evolutionary process has adapted the human eye to respond to wavelengths that correspond closely—but not exactly—to those received from the Sun. The difference probably arises from the fact that there is simply too much radiation in the infrared region to be accomodated by the optic system. The eye is sensitive to only about 14 percent of the solar radiation received at the Earth's surface.

Summary of Important Ideas

For a particle undergoing simple harmonic motion, the displacement is a *sinusoidal* function of the time.

The *period* of simple harmonic motion is independent of the amplitude of the vibration.

Waves of all types obey the principle of *superposition*.

Sound waves in gases are *longitudinal* waves; electromagnetic waves are *transverse* waves.

The observed *frequency* of a wave depends on the relative motion of the source and the observer (the *Doppler effect*).

Standing waves are produced on strings or in enclosures when the wave disturbance always has *nodes* at the end positions.

The propagation of a wave in space can be determined by considering every point on a given wave front to be the source of out-going spherical waves (*Huygens' principle*).

Diffraction causes waves to "bend around" obstacles and to spread out upon passing through narrow apertures.

When waves from two (or more) sources arrive at a particular point, *constructive interference* results if the waves are *in phase* and *destructive interference* results if the waves are *out of phase*.

Electromagnetic waves are produced only by electric charges that are undergoing *accelerations*.

In a *plane electromagnetic wave* the field vectors, **E** and **B**, are mutually perpendicular.

An electromagnetic wave possesses and can transfer both *energy* and *momentum*.

All electromagnetic waves have *identical* properties; they can differ only in *frequency*.

OSCILLATIONS,
WAVES, AND
RADIATION

Questions

8.1

Sound waves do not propagate for great distances through air; eventually, the waves "die out." What happens to the *energy* in the sound wave?

8.2

Explain why small rooms are not generally suited for listening to high-fidelity music.

8.3

What do you think would be your bodily reaction to an intense sound wave of wavelength $\lambda = 50$ m? (What is the frequency of the wave?)

8.4

Explain how the string in Fig. 8.6 (fifth frame), which has no displacement at any position, can have a displacement at a later time. Describe the physical situation that allows a displacement to be generated from a condition of no displacement.

8.5

Is there a Doppler effect when the source moves perpendicular to the line connecting the source and the observer?

8.6

Suppose that a double slit is illuminated with *white light* (that is, light containing *all* frequencies) instead of *monochromatic light* (light of a *single* frequency). Describe the interference pattern that would be produced.

8.7

A sound wave in air and an electromagnetic wave in air each have a wavelength of 10 cm. Classify the two waves (audible, inaudible; light, radio wave, etc.). Why are two waves of the same wavelength so different in their properties?

8.8

The Sun radiates huge amounts of electromagnetic energy. We know that such radiation carries momentum. Does the momentum of the Sun change with time? Explain.

8.9

Argue that if a particle in the interplanetary region of the solar system is sufficiently small, radiation pressure will overcome the solar gravitational attraction and the particle will be accelerated *away* from the Sun. (Consider the particle to be spherical. What properties of the particles determine the magnitude of the gravitational force? What properties determine the magnitude of the force exerted by radiation pressure? Examine the way in which these forces change as the size of the particle is reduced.) Particles with radii less than about 2×10^{-6} m are actually expelled from the solar system by radiation pressure.

8.10

Huge antennas are required to generate radio waves, but X rays are produced by atoms and gamma rays are produced by nuclei. Why are electromagnetic waves of the highest frequency generated by the smallest systems?

Problems

8.1

A 10-g mass attached to a spring vibrates with a period of 2 sec. How much force is required to stretch the spring by 10 cm starting from its equilibrium position? (Use the approximation that $\pi^2 \cong 10$.)

8.2

A flash of lightning is observed and 8 sec later a clap of thunder is heard. About how far away was the lightning?

8.3

A stick of dynamite is exploded on the surface of the sea. The sound is propagated through the water as well as the air. At a distance of 3 km, which signal will be heard first? What will be the time interval between the arrival of the two signals?

8.4

What is the wavelength of a 10-kHz sound wave in iron? What would be the wavelength of the same wave in air?

8.5

An organ note ($\lambda = 22$ ft) is sustained for 1 sec. How many full vibrations of the wave have been emitted?

8.6

Waves travel with a velocity of 20 m/sec on a certain taut string, which is attached to two supports that are separated by 2 m. What are the frequencies of the first four standing waves (starting with the longest wavelength) that can be set up in the string? Which of these modes of vibration will produce an audible sound?

8.7

Two narrow slits are so close together that a direct measurement of their separation is difficult to make. By illuminating these slits with light ($\lambda = 5 \times 10^{-7}$ m) it is found that on a screen 4 m away adjacent bright lines in the interference pattern are separated by 2 cm. What is the separation of the slits?

8.8

Yellow light ($\lambda = 6 \times 10^{-7}$ m) illuminates a single slit whose width is 0.1 mm. What is the distance between the two dark lines on either side of the central maximum if the diffraction pattern is viewed on a screen that is 1.5 m from the slit?

OSCILLATIONS,
WAVES, AND
RADIATION

8.9

Microwaves ($\lambda = 0.5$ cm) are incident on a pair of slits that are separated by 25 cm. Describe the intensity pattern that would be found by moving a microwave detector along a screen that is 15 m from the slits.

8.10

What is the wavelength of the 25-MHz radiation that WWV uses to broadcast time signals? What is the frequency of radiation in the 10-m shortwave broadcast band?

8.11

A proton oscillates back and forth across the diameter of a nucleus (about 10^{-14} m) with a velocity $v \cong 0.05c$. What is the approximate frequency of the emitted radiation? (The frequency of the radiation is equal to the frequency of the oscillation.) How would you classify this radiation?

CHAPTER 9
RELATIVITY

Maxwell's theory of electromagnetism was well established as the 19th century drew to a close, and it was understood that light is a wave phenomenon correctly described by Maxwell's equations. An integral part of the theory of the propagation of electromagnetic waves was the concept of the *ether*.[1] Because of the mechanistic view of electromagnetism that was in vogue, it was considered essential that an ether exist in order to provide a medium in which the waves could propagate. Maxwell's equations were considered valid in a reference frame at rest with respect to the ether. Unlike Newton's equations, which were known to be valid in *all* reference frames, Maxwell's equations seemed to demand a *preferred* reference frame.

Despite the efforts of many able physicists and mathematicians, the ether theory was never completely successful in the interpretation of the various experimental results regarding the propagation of light. In 1905, a crucial new idea was contributed by Albert Einstein (1879–1955). In a single bold stroke, Einstein swept away the ether theory with all its complications and replaced it with only two postulates. Using these postulates as a foundation, he was able to construct a beautiful theory which is a model of logical precision. Einstein's relativity theory provided the key link between mechanics and electromagnetism—it unified the two great theories of classical physics.

When we first encounter them, the ideas of relativity seem somewhat strange and forced. But relativistic effects are important only when velocities approaching the velocity of light are encountered, and our intuition is based on our everyday experience in which we almost never meet situations involving such high velocities. Perhaps if we were reared in a much faster-moving world, relativistic concepts would be natural and easy to accept. Nevertheless, if experimental facts conflict with our preconceived notions, we cannot change the facts—only our ideas. After all, it was this same brand of "common sense" that once supported the view that the Earth is flat.[2]

---

[1] See the introductory section of Chapter 6 for additional comments on the *ether*.
[2] Einstein once remarked that "common sense" is the layer of prejudices built up before the age of 18.

Fig. 9.1 *Albert Einstein as a young man (1905).*

9.1   *The Basis of Relativity*

EINSTEINS'S POSTULATES

Einstein's great triumph was the realization that it was possible to remove all of the apparent discrepancies between the dynamics of mechanical and electromagnetic systems by basing a new theory on only two postulates:

I. *All physical laws are the same in all inertial reference frames.*

II. *The velocity of light (in vacuum) is the same for any observer in an inertial reference frame regardless of the relative motion between the light source and the observer.*

The theory that is based on these postulates and that applies to all nonaccelerating systems is called the *special theory of relativity.* (The more complicated situation of accelerating systems is the subject of the *general theory of relativity,* which is described briefly in Section 9.5.) It is indeed remarkable that such a far-reaching theory—one that forced a complete reexamination of the traditional views concerning the fundamental concepts of space and time and that has had such a profound effect on the inter-pretation of atomic, nuclear, and astrophysical effects—can be built on only two postulates as simple as those given by Einstein. If it is the goal of physical theory to formulate the laws of Nature with brevity and economy of as-sumptions, then relativity theory is surely the showpiece of science.

The first of Einstein's postulates is not too difficult to accept. Indeed, Newton fully appreciated the fact that the laws of mechanics are the same for all inertial observers. (A ball tossed into the air will proceed straight up and straight down whether in a stationary or a moving automobile—if we are careful to view the motion in a reference frame fixed with respect to the automobile.) But the ether theory of electrodynamics demanded a *preferred* reference frame associated with the ether. Einstein realized that

RELATIVITY

he could eliminate the necessity for the ether and place *all* physical laws on the same basis by making an additional postulate that the velocity of light is the same for all observers. But how can we accept such a statement when it runs counter to our intuitive expectations?

THE VELOCITY OF LIGHT IS CONSTANT

Suppose that an airplane moves with a velocity $v = 400$ mi/hr with respect to a stationary Earth-based coordinate system (Fig. 9.2). If a gun mounted on the airplace fires a shell in the forward direction with a velocity $u = 600$ mi/hr *with respect to the airplane,* what will be the velocity of the shell as measured by a ground-based observer? According to our Newtonian ideas, the answer is simply $V = v + u = 1000$ mi/hr.

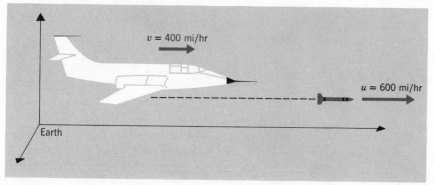

Fig. 9.2 *To an Earth-based observer the velocity of the shell is V = v + u = 100 mi/hr.*

Will there be any difference in the way in which we make our analysis if we increase the velocity of the airplane to one-half the velocity of light[3] and fire a light pulse instead of a shell? According to Einstein's second postulate, we cannot simply add the two velocities to obtain $0.5c + c = 1.5c$ for the velocity of the light pulse as measured by the observer on the ground. Einstein's statement is that the light pulse travels with the velocity $c = 3 \times 10^8$ m/sec with respect to the airplane *and* with respect to the ground! The velocity of the source (the airplane) does not influence the velocity of the light pulse as measured *by any observer* in an inertial reference frame.

This prediction can be put to experimental tests. A particularly elegant demonstration was recently made in the following way. A large accelerator was used to produce a beam of extremely high speed neutral π mesons (π° mesons or neutral *pions*). These π° particles have the property that they decay spontaneously into energetic photons (or *gamma rays*). The velocity of the particles in the beam was measured to be $0.99975c$, and when the decays took place, the velocity of the photons was also measured. The result was that the photons traveled with a velocity that was equal to c within

[3]Modern experiments have shown that the velocity of light (in a vacuum) is $c = (2.997\,925 \pm 0.000\,010) \times 10^8$ m/sec. In this book, we shall usually use the approximate value, $c = 3 \times 10^8$ m/sec.

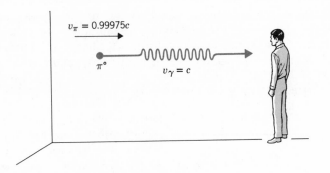

Fig. 9.3 *Neutral pions (π°) traveling with a velocity $v_\pi = 0.99975c$ decay into gamma rays that travel with a velocity $v_\gamma = c$ with respect to the laboratory observer.*

1 part in 10^4 (see Fig. 9.3). This is a direct and striking confirmation of Einstein's postulate.

In the following section we will examine some of the consequences of Einstein's remarkable postulates.

9.2   *Time and Length in Special Relativity*

SIMULTANEITY

In our everyday experience we have become accustomed to consider that all events proceed in time in an orderly and regular way—there is a past, a present, and a future, and we can always establish whether one event preceded or followed another event or whether the two events occurred simultaneously. Einstein showed, however, that in the relativistic world events that appear to occur in a certain sequence according to one observer may appear to occur in quite a different sequence to another observer who is in motion with respect to the first observer. This is perhaps the most startling result of the Einstein theory, but it is easy to show that this conclusion follows in a direct and simple way from the constancy of the velocity of light.

In order to demonstrate that time is a relative concept, consider the following example (which is due to Einstein). In Fig. 9.4*a* an observer K sees two lightning bolts strike the ends of a moving railway car just as the midpoint of the car passes him. Since the ends of the car are equidistant from him, K sees the light from the flashes *simultaneously*. Observer K' stands in the middle of the car. Now, K knows that K' is moving toward the flash of light that originates at B and *away* from the flash that originates at A. Therefore, K concludes that the B flash will reach K' *before* the A flash reaches K'. But K' is a stationary observer in an inertial reference frame (the railway car) and he knows that both flashes of light travel with a velocity c in his reference frame. Since K' is equidistant from the two ends of the car and since the flash from B reaches him first (Fig. 9.4*b*), he concludes that the B flash must have occurred *before* the A flash. Thus, two events that appear to be simultaneous in the K system do *not* appear to be simultaneous in the K' system because the two systems are in relative motion.

If K had seen the lightning strike A *slightly* before the strike at B occurred, he would have seen A precede B, whereas K' would have still have seen

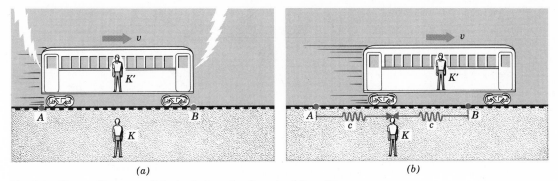

Fig. 9.4 *Observer* K *sees two lightning bolts strike the ends of the railway car simultaneously. But observer* K', *who is moving toward the right-hand bolt with a velocity* v, *sees the right-hand bolt strike first.*

B precede *A*. Hence, the two observers would have seen the *opposite* time ordering of events; "past" and "future" would have been interchanged.

Although the time sequence of events as seen by different observers depends on their relative velocity, the physical law of *cause and effect* must still be valid in the realtivistic world; *no observer (whatever his motion) can perceive an event which is an effect prior to an event which is the cause of the first event.*

TIME DILATION

In what way does the fact that two observers are in relative motion affect their measurements of time intervals? In order to examine this question we must have an appropriate clock. Let us prepare a "standard clock" in the following way. At a distance L from the origin along the y-axis we place a mirror M, as in Fig. 9.5a. At the origin we place a light flasher and a light detector. The standard unit of time will be the interval required for light to travel from the flasher to the mirror and back to the detector. If the K observer operates his flasher at $t = 0$, he finds that the light pulse makes the round trip and returns to the origin at the time.

$$t = \frac{2L}{c} \tag{9.1}$$

Now, an identical clock is installed in the K' system and the K *observer views its operation.* The K' system moves past the K system at a velocity v along the x-axis. At the instant when O and O' coincide ($t' = 0$), the flasher of the K' clock is operated. Since the K' system is moving relative to K, the K observer notices that the light flash must travel from O' to M' on a slanted path that is longer than the path traveled by the light flash in the K clock. When the light flash reaches M', this time corresponds to one half of a standard interval of the K' clock, that is, $\frac{1}{2}t'$ (Fig. 9.5b). The complete standard interval t' ends when the reflected light flash again reaches O' having traveled the path OPO' (Fig. 9.5c). During this time, the origin O' has moved a distance vt' from O'. We can see that the K observer believes that the K' clock runs slowly because the light signal moves along a path which is *longer* than the path for the light pulse in the K clock.

9.2
TIME AND
LENGTH IN
SPECIAL
RELATIVITY

197

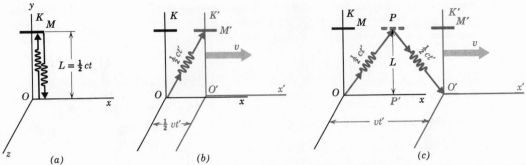

Fig. 9.5 *According to the observer in* K, *it requires a shorter time for the round trip* OMO *of the light signal in* K *than for the round trip* OPO′ *in* K′. *The* K *observer concludes that the clock in* K′ *runs* slower *than the clock in* K.

In order to compare the intervals t and t', we use the Pythagorean theorem for the triangle $OM'O'$ (Fig. 9.5b) or OPP' (Fig. 9.5c). Thus,

$$(\tfrac{1}{2}ct')^2 = (\tfrac{1}{2}vt')^2 + L^2$$

or, using Eq. 9.1 for L,

$$(\tfrac{1}{2}ct')^2 = (\tfrac{1}{2}vt')^2 + (\tfrac{1}{2}ct)^2$$

cancelling the factor $(\tfrac{1}{2})^2$ and transposing the term $(vt')^2$, we find

$$(c^2 - v^2)t'^2 = c^2t^2$$

so that

$$t'^2 = \frac{c^2t^2}{c^2 - v^2} = \frac{t^2}{1 - \dfrac{v^2}{c^2}}$$

or finally,

$$t' = \frac{t}{\sqrt{1 - \dfrac{v^2}{c^2}}} = \frac{t}{\sqrt{1 - \beta^2}} \tag{9.2}$$

where we have used the customary notation, $\beta = v/c$.

The standard time interval t' of the K' clock, *as viewed by the K observer*, is *longer* than the interval of the K clock. (The K observer must draw this conclusion since he views the light flash in the K' clock traveling a greater distance than the light flash travels in his own clock.) Therefore, the observer in K finds that the K' clock runs *slower* than his clock. Of course, if the K' observer views the K clock, he concludes that the K clock runs slower than his. Therefore, we can state that *any observer will find that a moving clock runs slower than an identical clock that is stationary in his reference frame*.

We used a light-flash clock for this argument; will the same result be found if we use some other type of clock, such as a mechanical clock? The answer is that indeed it must. Consider the alternative. Suppose that we have

RELATIVITY

a clock that does not slow down when in motion relative to our particular reference frame. We could then set this clock to agree with our standard clock and then (at least in principle) send this clock on an expedition to all manner of moving reference frames for the purpose of adjusting their clocks to agree with ours. By using these synchronized clocks we would then be able to determine unambiguously the time ordering of events as viewed in any reference frame. But we have already concluded that the finite velocity of light prevents any such determination of the absolute sequence of events in moving systems. We are therefore forced to concede that it is impossible to construct such a "perfect" clock.

TIME DILATION IN PION DECAY

We can illustrate the time dilation (or time *expansion*) effect by considering the motion of the short-lived pions. When viewed at rest, pions have an average lifetime of $\tau_\pi = 2.6 \times 10^{-8}$ sec before decaying into other elementary particles. Pions are produced in copious quantities by the interaction of high-energy protons with matter and therefore are relatively easy to study.

If pions move with a velocity of $0.75c$, the average distance they would travel before decay is $l_\pi = v\tau_\pi = 0.75 \times (3 \times 10^8$ m/sec$) \times (2.6 \times 10^{-8}$ sec$) = 5.85$ m. At the Columbia University cyclotron a beam of pions with $v = 0.75c$ was produced and it was found that the average distance these particles traveled before decay was not 5.85 m but 8.5 ± 0.6 m. We can account for this difference in terms of the time dilation effect. Because the pions are moving in the laboratory system (corresponding to our K system), the laboratory observer sees any clock in the system moving with the pions (the K' system) running slowly. But the decay rate of the pions is a type of clock and so the laboratory observer will find the average lifetime of the pions to be longer than τ_π. In fact,

$$\tau_{\text{lab}} = \frac{\tau_\pi}{\sqrt{1 - \beta^2}} = \frac{2.6 \times 10^{-8} \text{ sec}}{\sqrt{1 - (0.75)^2}} = 3.9 \times 10^{-8} \text{ sec}$$

Therefore, the average distance in the laboratory that the pions will travel before decay is

$$l_{\text{lab}} = v\tau_{\text{lab}} = 0.75 \times (3 \times 10^8 \text{ m/sec}) \times (3.9 \times 10^{-8} \text{ sec}) = 8.8 \text{ m}$$

which agrees with the measured value of 8.5 ± 0.6 m.

LENGTH CONTRACTION

Let us now view the decay of the pions from the standpoint of an observer moving with the pions. In the pion rest-frame, the pions will live an average of 2.6×10^{-8} sec and will travel an average distance of 5.85 m. But we know that the average distance *as measured in the laboratory* is 8.8 m. The distance through which the pions moved before decay is the *same physical* distance whether measured in the pion frame or in the laboratory frame.

We can only conclude that not only are *time* measurements different in two reference systems in relative motion but so are *distance* measurements. The observer in the pion frame sees the laboratory moving past him with a velocity of $0.75 c$. The length of 8.8 m, as measured by an observer in

9.2
TIME AND
LENGTH IN
SPECIAL
RELATIVITY

199

the laboratory frame, appears to the observer in the pion frame to be only 5.85 m. That is, *there is a contraction of length due to relative motion.* The length l' of an object in motion relative to an observer is contracted compared to the length l of an identical object at rest with respect to the observer:

$$\boxed{l' = l\sqrt{1 - \beta^2}} \tag{9.3}$$

A measurement made by the observer in the pion frame of the 8.8 m laboratory length gives $l' = (8.8\ \text{m}) \times \sqrt{1 - \beta^2} = 5.85$ m.

It should be noted that the contraction of length takes place only *along* the direction of relative motion; the dimensions of an object *transverse* to the direction of motion are unaffected. (This is why the mirror in the light clock of Fig. 9.5 was placed on the y-axis and not on the x-axis; the observers in both the K and K' systems measure the same distance L for the distance from the origin to the mirror.)

Are the time dilation and length contraction effects *real?* The only possible answer to such a question is that an effect is real if and only if it can be measured. All measurements made in systems that are in relative motion confirm the time dilation and length contraction effects.

Example **9.1**

An observer moves past a meter stick with a velocity that is one-half the velocity of light. What length does he measure for the meter stick?

$$l' = l\sqrt{1 - \beta^2} = (1\ \text{m}) \times \sqrt{1 - (0.5)^2}$$
$$= (1\ \text{m}) \times \sqrt{0.75} = 0.866\ \text{m}$$

In order to obtain numerical results for problems that involve the relativistic factor $\sqrt{1 - \beta^2}$; the following approximate expressions can be used when the velocity v is very small compared to c:

$$\text{If } \beta \ll 1\ (v \ll c)\text{: } \sqrt{1 - \beta^2} \cong 1 - \tfrac{1}{2}\beta^2 \tag{9.4a}$$

$$\frac{1}{\sqrt{1 - \beta^2}} \cong 1 + \tfrac{1}{2}\beta^2 \tag{9.4b}$$

Example **9.2**

Suppose that the velocity of the observer relative to the meter stick in the previous example is reduced to 30 m/sec (about 67 mi/hr). What length does he now measure for the meter stick?

$$l' = (1\ \text{m}) \times \sqrt{1 - \left(\frac{30\ \text{m/sec}}{3 \times 10^8\ \text{m}}\right)^2}$$
$$= (1\ \text{m}) \times \sqrt{1 - 10^{-14}}$$
$$\cong (1\ \text{m}) \times (1 - 0.5 \times 10^{-14})$$
$$= 0.999999999999995\ \text{m}$$

It is easy to see that the relativistic contraction of length is of little practical consequence in everyday matters!

THE TWIN PARADOX

One of the results of relativity theory that has been much discussed (and misunderstood) in recent years is the so-called "twin-paradox." Suppose that there are twins, Al and Bob, and that Bob is an astronaut. Bob embarks on a space journey to a star that is 10 light years distant; Al remains on Earth. If Bob's space ship travels at a velocity of $0.99c$ relative to the Earth, according to Al the trip will require a time[4]

$$\Delta t = \frac{10 \text{ L.Y.}}{0.99c} \cong 10 \text{ years}$$

An equal time will be required for the return journey, so Bob will arrive back on Earth when Al is 20 years older than when Bob departed.

In Bob's space ship, however, the Earth and the star appear to be moving with a velocity of $0.99c$ relative to Bob. Therefore, the Earth-star distance is contracted to

$$l' = (10 \text{ L.Y.}) \times \sqrt{1 - (0.99)^2} = 1.4 \text{ L.Y.}$$

According to Bob's clock, the trip will require only 1.4 years and he will return to Earth after having aged by 2.8 years. When he again greets his brother, Bob discovers that his twin is $20 - 2.8 = 17.2$ years *older* than he is! But we know that all motion is relative. Therefore, if the trip is viewed from Bob's reference frame, he sees Al (and the Earth) go on a round trip journey. Hence, Al's clock should run more slowly than Bob's and when Al returns (along with the Earth), Bob should find that his twin is *younger* than he is. Thus, the paradox.

The "paradox" rests on invoking the symmetry of the situation. It should not matter which twin takes the trip and which remains at home. But it *does* matter, because *Al* (the stay-at-home) *is always in an inertial reference frame whereas Bob* (the traveler) *has undergone accelerations.* In leaving the Earth, Bob was accelerated to $0.99c$; he was accelerated when he turned around at the star; and he was accelerated again when he returned to Earth and landed. Therefore, the situation is *not* symmetric between Al and Bob. Because inertial reference frames are not involved throughout, the analysis must be carried out quite carefully. A proper calculation (which can be made within the context of special relativity if appropriate care is exercised) does in fact show that Bob ages less rapidly than his twin.

Because of the time dilation effect, we can imagine the exciting possibility of traveling to distant stars. If the trip is made at a velocity sufficiently close to the velocity of light, the traveler can easily cross vast distances of space within a time short compared to his lifetime. But he would return to a different Earth—one that has progressed (?) by hundreds or even thousands of years during his absence. There is, of course, a difficulty in this fanciful picture; we now have absolutely no conception of how to generate sufficient energy to accelerate a space ship to velocities that approach c!

It should be emphasized that the "twin paradox" is a real effect; the traveling twin will age less rapidly than his Earth-bound brother. On the other hand, the traveler cannot take advantage of his longevity because all

[4]Since 1 L.Y. is the distance traveled by light in 1 year, the quantity 1 L.Y./c is just *1 year*.

of his biological processes progress at a slower rate (compared to the Earth rate) and he must function, think, and perform at this reduced pace.

THE VELOCITY ADDITION RULE

How can the statement that the velocity of light is independent of the motion of the source be consistent with the fact that all ordinary mechanical velocities simply *add* algebraically, as in Fig. 9.2? Einstein showed that the simple addition formula for mechanical velocities is not correct and must be modified. If two velocities v_1 and v_2 are to be added, the sum is

$$V = \frac{v_1 + v_2}{1 + \dfrac{v_1 v_2}{c^2}} \qquad (9.5)$$

In the event that v_1 and v_2 are small compared to the velocity of light (the case for all ordinary mechanical situations), the term $v_1 v_2/c^2$ is much less than unity and can be neglected. Then, the sum velocity is $V = v_1 + v_2$, identical to the result that would be obtained by applying Newtonian reasoning.

If one of the velocities is the velocity of light, $v_1 = c$, then

$$V = \frac{c + v_2}{1 + \dfrac{cv_2}{c^2}} = \frac{c + v_2}{1 + \dfrac{v_2}{c}} = \frac{c + v_2}{\left(\dfrac{c + v_2}{c}\right)} = c$$

This result insures that the velocity of light is the same for all observers because no matter what velocity v_2 is added to c, the addition rule always yields c. In particular, if $v_1 = c$ and $v_2 = c$, we still have $V = c$.

Example **9.3**

A spaceship moving away from the Earth at a velocity $v_1 = 0.75\,c$ with respect to the Earth, launches a rocket (in the direction *away* from the Earth) that attains a velocity $v_2 = 0.75\,c$ with respect to the spaceship. What is the velocity of the rocket with respect to the Earth?

$$V = \frac{v_1 + v_2}{1 + \dfrac{v_1 v_2}{c^2}} = \frac{0.75c + 0.75c}{1 + \dfrac{(0.75c)(0.75c)}{c^2}} = \frac{1.5c}{1 + 0.5625} = 0.96c$$

Therefore, in spite of the fact that the simple sum of the two velocities exceeds c, the actual velocity relative to the Earth is slightly less than c.

9.3 *Variation of Mass with Velocity*

A COLLISION EXPERIMENT

The first postulate of relativity theory is that all physical laws must be the same in all inertial reference frames. One of these laws is the conservation of linear momentum and we shall now make use of the invariability of this law to assess the effect of motion on mass.

Consider two observers who are stationed in two reference frames, K and K', that are in relative motion with the velocity v, as in Fig. 9.6. In each reference frame there is a stationary mass m_0. (That the two masses are in fact identical can be established beforehand by a balance comparison when the masses are at rest relative to one another.) The positions of the masses

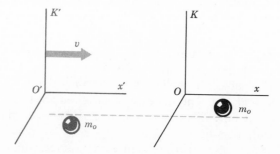

Fig. 9.6 *A mass* m_0 *is stationary in each reference frame. The relative velocity is* v.

are such that when the reference frames pass one another, a grazing collision of the masses takes place. That is, each mass receives a small velocity at right angles (that is, *transverse*) to the direction of the relative velocity of K and K'. (In such a collision neither mass will receive any appreciable longitudinal velocity relative to its own reference frame.) Therefore, after collision, the situation is that shown in Fig. 9.7. The mass in K has a

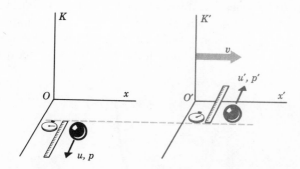

Fig. 9.7 *The grazing collision between the masses produces for each mass a velocity and a momentum transverse to the direction of relative motion.*

transverse velocity u and a transverse momentum p as measured by the observer in K; similarly for the mass in K' (but with primed quantities). Each observer uses a meter stick and a clock to measure the transverse velocity of his mass in his reference frame. Each observer obtains a numerical result for the velocity of his mass, which result he then communicates to his colleague in the other reference frame. They are both happy to note that the results are identical and congratulate themselves on having verified the conservation of linear momentum in the collision. In order to check the results, they decide to repeat the experiment twice again—once, K will observe K' making his measurements, and then K' will observe K making his measurements.

On the first rerun, K confirms that the meter stick used by K' is properly calibrated (transverse dimensions are unaffected by relative motion; length contraction takes place only *along* the direction of relative motion), but that his clock runs *slowly*. Therefore, when K' reported that his mass traveled 1 meter in T seconds, K concludes that by *his* clock it required a time *greater* than T seconds for the 1-meter trip. Thus, K calculates that the velocity of the mass in K' is *smaller* than the value u' reported by K'—in fact, smaller by the time-dilation factor $\sqrt{1 - \beta^2}$. If the velocity is smaller and if conservation of linear momentum is still to hold, then the mass used by K' must be (so argues K) *larger* than that used by K—in fact, larger by the amount $1/\sqrt{1 - \beta^2}$.

Of course, during the second rerun of the experiment, K' draws exactly the same conclusions about the measurements of K. *Both* observers therefore agree that the mass of an object in motion is greater than the mass of an identical object which is at rest. The increase of mass with velocity (just as length contraction and time dilation) is symmetrical between the two reference frames in relative motion.

The mass of an object as measured in a reference frame at rest with respect to the object is denoted by m_0 and is called the *rest mass* or *proper mass*. Then, the mass m as measured by an observer moving with a velocity v relative to the object is

$$m = \frac{m_0}{\sqrt{1 - \beta^2}} \qquad (9.6)$$

We must conclude from this equation that no material particle can attain or exceed the velocity of light because if $v = c$, the term $\sqrt{1 - \beta^2}$ vanishes and m becomes infinite. An infinite mass is a meaningless concept and we are therefore forced to accept the conclusion that material particles *always* move with velocities that are less than the velocity of light. And, because

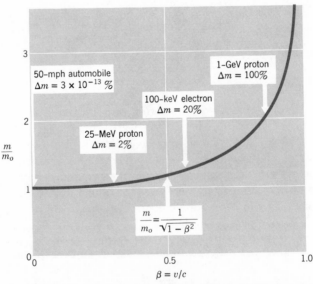

Fig. 9.8 *The relativistic increase of mass with velocity. The energies given for the elementary particles are the kinetic energies.*

of the velocity addition rule (Eq. 9.5), the velocity of a material particle is less than c in *any* reference frame.

PARTICLES WITH VARYING MASS

The difference between the mass m and the rest mass m_o is quite small unless the relative velocity v is greater than a few percent of the velocity of light. Therefore, the relativistic increase of mass with velocity is undetectable for all everyday velocities; it is only when we deal with elementary particles that have been given high velocities in accelerators that we encounter appreciable mass increases. Figure 9.8 shows Eq. 9.6 in graphical form; indicated on the curve are points corresponding to one everyday object (an automobile traveling at 50 mi/hr) and three high-velocity elementary particles. Notice that a 1-GeV proton has *twice* the mass of a proton at rest, whereas the mass increase of a 50-mi/hr automobile is insignificantly small.

9.4 Mass and Energy

EINSTEIN'S MASS-ENERGY RELATION

If $v \ll c$, we can use the approximation given in Eq. 9.4b and then Eq. 9.6 can be expressed as

$$m \cong m_o(1 + \tfrac{1}{2}\beta^2) \qquad (v \ll c)$$

Multiplying both sides of this equation by c^2 and noting that $c^2\beta^2 = v^2$, we find

$$mc^2 \cong m_o c^2 + \tfrac{1}{2}m_o v^2 \qquad (v \ll c) \tag{9.7}$$

The term $\tfrac{1}{2}m_o v^2$ is just the Newtonian result for the *kinetic energy*. The term $m_o c^2$ is clearly some *intrinsic* aspect of the object because it depends only on the *rest* mass. We call this quantity the *rest energy* of the object. The sum of the *rest* energy and the *moving* energy (that is, the kinetic energy) is the *total energy* of the object:

$$\underset{\text{(Total energy)}}{mc^2} = \underset{\text{(Rest energy)}}{m_o c^2} + \underset{\text{(Kinetic energy)}}{KE} \tag{9.8}$$

This equation is just the expression of the Einstein mass-energy relation:

$$\boxed{\mathcal{E} = mc^2} \tag{9.9}$$

where $\mathcal{E}$ is the *total* energy (rest energy + kinetic energy) of the object.

Although we used the approximation $v \ll c$ in order to obtain Eq. 9.7, the final equations which we have obtained, Eqs. 9.8 and 9.9, are nevertheless *exact*.

Example **9.4**

What is the mass of an electron that has a kinetic energy of 2 MeV? First, we calculate the rest energy of an electron:

$$m_o c^2 = (9.11 \times 10^{-31} \, \text{kg}) \times (3 \times 10^8 \, \text{m/sec})^2$$

$$= (8.2 \times 10^{-14} \, \text{J}) \times \left(\frac{1 \, \text{MeV}}{1.6 \times 10^{-13} \, \text{J}} \right)$$

$$= 0.511 \, \text{MeV}$$

The kinetic energy expressed in units of $m_o c^2$ is

$$KE = 2 \, \text{MeV} = (2 \, \text{MeV}) \times \left(\frac{m_o c^2}{0.511 \, \text{MeV}} \right) \cong 4 \, m_o c^2$$

Therefore,

$$mc^2 = m_o c^2 + KE \cong m_o c^2 + 4 \, m_o c^2 = 5 \, m_o c^2$$

Hence, the mass of a 2-MeV electron is approximately 5 times the mass of an electron at rest.

Energy and mass are related according to the expression in Eq. 9.9, but it is *not* correct to say that this means that energy and mass are the same thing; energy and mass are distinct physical concepts. Einstein's mass-energy relation *does* say that energy *has* mass (*not* "energy *is* mass").

9.5 *General Relativity*

THE PRINCIPLE OF EQUIVALENCE

Thus far we have considered only motions that take place with constant velocity. In the *general theory of relativity* we include considerations of gravitational fields and therefore *accelerations*. In fact, the general theory is a theory of *gravitation*.

The first important aspect of the general theory has to do with the equivalence of gravitational fields and accelerated motion. If we are in a laboratory on the Earth, as in Fig. 9.9a, a mass that is released will accelerate downward due to the gravitational attraction of the Earth. Now, let us move this laboratory into space, away from the gravitational influence of the Earth or any other body, and attach it to a rocket that is accelerating, as in Fig. 9.9b. If the magnitude of the rocket's acceleration a is equal to the acceleration due to gravity g and if the rocket pushes on the floor of the laboratory, the floor will be accelerated toward a mass that is released. Insofar as observations of the motion of the mass relative to the floor are concerned, the accelerated motion in the two cases will be exactly the same. If the laboratory has no windows, the observer can never distinguish between an acceleration due to gravity and an acceleration due to a push by a rocket.

Einstein incorporated this reasoning into his general theory by postulating the *principle of equivalence*:

In a closed laboratory, no experiment can be performed that will distinguish between the effects of a gravitational field and the effects due to an acceleration with respect to an inertial reference frame.

An immediate consequence of this postulate is that the mass of an object involved in a gravity experiment (the *gravitational* mass, Fig. 9.9a) is exactly

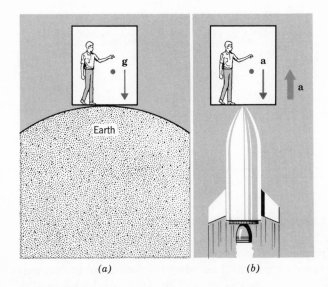

Fig. 9.9 *If* a = g, *the observer cannot distinguish between the acceleration produced by gravity and that produced by the push of the rocket. (The observer's box laboratory has no windows.)*

equal to the mass of the same object involved in an acceleration experiment which is not influenced in any way by gravity; this latter mass is called the *inertial* mass (Fig. 9.9b). Why should the gravitational mass of an object be equal to its inertial mass? This is one of the fundamental unanswered questions in physics today. In the general theory, the equivalence principle is simply a postulate. But experimental tests can be made. Recent measurements have shown that if there is any difference between gravitational mass and inertial mass, it must be less than 1 part in 10^{11}. In spite of the high precision of this result, an even sharper confirmation of the assertion of the equivalence principle would be desirable in order to provide greater support for this fundamental postulate of the general theory.

TESTS OF THE GENERAL THEORY

In Section 4.2 it was mentioned that the perihelion of Mercury's orbit has been observed to move in space (that is, to *precess*) at a rate that is larger than that predicted on the basis of Newtonian dynamics. After subtraction of the calculable perturbations due to the other planets, there remains a net precession of 43.11 ± 0.45 seconds of arc per century. Einstein was able to obtain, on the basis of his general theory, the value 43.03 seconds of arc per century. The excellent agreement between the calculated and the observed values is the most outstanding success of the general theory.

The general theory also predicts that a light ray passing close to a massive object will follow a path that is slightly curved. We can understand this result qualitatively if we recall that electromagnetic radiation, including light, has *energy* and that energy has equivalent *mass*. Therefore, a gravitational field will affect a light ray and cause it to be bent in much the same way that a fast particle would bend when passing by a massive object. Because light travels at such an enormous velocity there is only a brief time during which the "attraction" is effective, and hence the deflection is small even for a passage near such a massive object as the Sun.

This prediction can be tested by observing the shift in the apparent

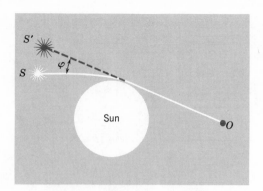

Fig. 9.10 *The light from a star* S *is bent upon passing close to the Sun, thus causing a shift in the apparent position of the star to* S′.

position of a star when its light passes close to the Sun, as shown in Fig. 9.10. Because of the brightness of the Sun, measurements of the effect are carried out by comparing the apparent positions of stars during a solar eclipse with the positions six months later when the Sun is not in that part of the sky and ordinary night photographs of the stars can be taken. The apparent shifts in position of hundreds of stars have been made by this technique and the average result is approximately 2 seconds of arc for the deflection; the general theory predicts 1.75 seconds of arc. Unfortunately, the uncertainty in the measurements is about 10 percent and there are some conflicting results, so that we cannot view this test as definitive.

New experiments are planned for the near future that will permit measurements to be made under daylight conditions, obviating the necessity of waiting for eclipses. These new measurements should decrease the uncertainty to about 1 percent and therefore will provide a stringent test of the general theory.

GRAVITATIONAL WAVES

In the previous chapter we found that an accelerating electric charge produces electromagnetic radiation. By analogy, then, should an accelerating massive object produce *gravitational* radiation? According to the general theory, the answer is *yes,* but the amount of energy radiated is so small that extraordinarily delicate equipment is required to detect even the radiation from an exploding star (a *nova* or *supernova*). Nevertheless, experiments of this type are being conducted with equipment of the type shown in Fig. 9.11, and current results indicate the presence of signals distinct from local effects (such as earthquakes) that would disturb the apparatus. It appears, therefore, that the general theory has been confirmed in another important respect—the prediction of the existence of gravitational waves.

GEOMETRY AND GRAVITY

Consider a pair of physicists who are *two-dimensional* men. That is, they appreciate length and breadth but have no comprehension of height. These physicists operate in a world that to them is a *plane.* Suppose these men are on the surface of the Earth at the equator—positions *A* and *B* in Fig. 9.12. They start out on a journey that takes them due north in parallel paths.

RELATIVITY

Fig. 9.11 *The gravitational wave detection apparatus of Professor Joseph Weber and his colleagues at the University of Maryland. The heart of the system is a 1400-kg cylinder of aluminum which is suspended in vacuum in a tank. Sensitive instruments can detect motion of the end of the cylinder (caused by a gravitational wave) corresponding to an average displacement of only 10^{-16} m!*

Professor J. Weber of University of Maryland

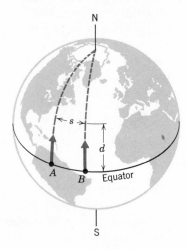

Fig. 9.12 *The two-dimensional physicists start from* A *and* B *and travel northward in parallel paths. They conclude that they are attracted together by some force.*

After traveling a certain distance d, they discover that their separation s is less than when they started. They conclude that they have been attracted together by some "force" and they give this "force" a name—*gravity*.

But, of course, there is no "force"—they have been deceived by the fact that their geometry is *curved* whereas they had assumed only plane Euclidean geometry in describing their positions. Thus it is with our real world. If we insist that the Universe be described by Euclidean geometry, then there is a mysterious force—gravity—for which we have no fundamental explanation. However, all of the effects of gravity result, in the general theory of relativity, from the non-Euclidean nature of the geometry (a four-dimensional space-time geometry) of the Universe. The presence of matter distorts the geometry and manifests itself in terms of the "gravitational force."

Einstein died (in 1955) still searching after his unproven vision that not only gravitation but all of the physical Universe can be completely described in terms of geometry alone. Current research by the disciples of Einstein has not succeeded in establishing this view of the Universe but the efforts continue.

Summary of Important Ideas

The *special theory of relativity* rests on two postulates:

I. The laws of physics are the same in all inertial reference frames.
II. The velocity of light in vacuum is the same for all observers in inertial reference frames.

In formulating the special theory of relativity, Einstein abandoned three basic ideas of Newtonian theory: (1) the concept of absolute space and time, (2) the principle of the addition of velocities, and (3) the law of conservation of mass (which he generalized to the conservation of mass-energy).

No *material particle* can have a velocity with respect to any inertial reference frame that is equal to or greater than the velocity of light; no *signal* can be transmitted at a velocity greater than c.

The *time sequence of events* as perceived by two observers depends on the relative motion of the observers. But no observer, regardless of his motion, can perceive an *effect* before he perceives the *cause* of that effect.

An observer who is in motion relative to a clock will find that clock to run more *slowly* than an identical clock at rest in his frame of reference (*time dilation*).

A measurement by an observer of the length of an object that is in motion relative to the observer will give a *smaller* result than a measurement carried out by an observer at rest with respect to the object (*length contraction*). (The contraction is only in the dimension *along* the direction of relative motion; the transverse dimensions are unaffected.)

A particle moving with respect to an observer has a mass *greater* than that of an identical object at rest with respect to the observer.

The *total energy* of an object is the sum of its *rest energy* m_0c^2 and its *kinetic energy;* this total energy is $\mathcal{E} = mc^2$.

The *principle of equivalence* states that the effects of gravity and accelerated motion are indistinguishable.

The predictions of the *general theory of relativity* have been verified for (a) the precession of the perihelion of Mercury's orbit, (b) the bending of light rays that pass near the Sun, and (c) the existence of gravitational waves. None of these tests is definitive, but the techniques used are being improved and new measurements are to be made in the near future.

Questions

9.1

The "writing speed" of a cathode-ray oscilloscope is the speed with which an electron beam can trace a line on the screen. A certain manufacturer claims that the writing speed of his oscilloscopes is 6×10^8 m/sec? Can his claim be true? Explain.

9.2

Using the results of relativity theory, comment on the anonymous limerick:

> There was a young lady named Bright,
> Who could travel much faster than light.
> She departed one day,
> In a relative way,
> And returned on the previous night.

9.3

Suppose that the velocity of light were suddenly to become 30 mi/hr. Describe a few of the effects this would have on everyday events.

9.4

Discuss the implications for relativity theory if the velocity of light were infinite.

9.5

A bicycle rider pedals past you at a velocity of 2.5×10^8 m/sec. Make a sketch of the way the bicycle would appear to you. Would the rider think you were your usual self?

9.6

Imagine a shaft drilled into the Earth and reaching to a depth of 1000 mi. Suppose that an observer in a closed box falls freely down this shaft. Argue that the principle of equivalence does not apply to this case. (Is the gravitational field *uniform?* What difference does this make?)

9.7

Should binary stars radiate gravitational waves? Why?

Problems

9.1

A man sets up equipment to detonate dynamite charges by sending electrical signals along a wire. (The signals travel with the velocity of light.) 1 μsec after he pushes his firing button, a charge explodes at a distance of 400 m. After an additional 1 μsec, a second charge explodes at a distance of 500 m. Could the pushing of the button have been responsible for both explosions?

9.2

The star nearest the Earth is *Proxima Centauri* (one of the three stars in the Alpha Centauri cluster); the distance is approximately 4.3 light years. If a space traveler were to make the trip from Earth to Proxima Centauri at a uniform speed of $v = 0.95 c$, how long would it take according to an Earth clock? How long would it take according to the space travelers clock?

9.3

An observer measures the length of a moving meter stick and finds a value of 0.5 m. How fast did the meter stick move past the observer?

9.4

A highway billboard is in the form of a square, 5 m on a side, and stands parallel to the highway. If a traveler passes the billboard at a speed of 2×10^8 m/sec, what will be the dimensions of the billboard as viewed by the traveler?

9.5

An observer sees one spaceship moving away from him at a velocity 0.9 c and another spaceship moving in the *opposite* direction at the same velocity. What does he conclude is the relative velocity of the two spaceships? (Consider this carefully; the Einstein velocity addition rule is *not* involved in the answer. Why?) Does this result agree with that obtained by the occupants of the spaceships? Why?

9.6

What is the velocity of a particle that has a kinetic energy equal to its rest energy?

9.7

A meter stick is in motion in the direction along its length with a velocity sufficient to increase its mass to twice the rest mass. What is the apparent length of the meter stick?

CHAPTER 10
THE FOUNDATIONS OF
QUANTUM THEORY

At the close of the 19th century many scientists viewed physics as a closed subject. The laws of mechanics and the theory of universal gravitation had been established for more than 200 years. Maxwell's theory of electromagnetism was complete. It was understood that matter consists of atoms. Thermodynamics had recently been placed on a firm foundation with the development of the statistical approach to systems with large numbers of particles. The great conservation principles—of energy, linear momentum, angular momentum, mass, and electrical charge—were well established and appreciated. What else of *real* importance could be discovered?

In spite of the general complacency among 19th-century physicists concerning the status of the subject, nevertheless, there *were* problems lurking about. It was soon to become clear that these were not all trivial problems and that the very heart of 19th-century physics was coming under violent attack. First, it was Einstein's relativity theory that forced a new way of thinking about such fundamental concepts as space and time. Even before this revolution could be digested, new and equally far-reaching questions were being asked about the nature of radiation and matter, how did they differ and in what ways were they the same—what was the inner structure of atoms—what was the origin of the newly discovered *radioactivity?* The answers to these questions began to emerge in the early years of the 20th century and culminated with the development of modern *quantum theory*.

10.1 *Electrons and Quanta*

THOMSON'S EXPERIMENTS

In 1897 Sir Joseph John Thomson (1856–1940) published the results of a series of experiments designed to study the nature of the particles that carry electrical current in evacuated discharge tubes (such as the tubes used in neon lights). By deflecting these particles in electric and magnetic fields, Thomson found that the particles carry a *negative* charge. Furthermore, he was able to measure the charge-to-mass ratio, e/m, obtaining a value of 1.2×10^{11} C/kg. These particles, whose mass is much smaller than the mass of any other microscopic particle, are simply *electrons*, the particles that constitute the outer portion of all atoms. Modern experiments have shown that the charge-to-mass ratio for electrons is not far from the result of Thomson's early measurements (see Table 10.1).

Thomson's discovery of the electron was the opening episode in the series of developments that has led to our modern ideas concerning the structure of matter.

Table **10.1** *Properties of the Electron*

Charge, e	1.6022×10^{-19} C
Mass, m_e	9.1096×10^{-31} kg
Rest-mass energy, $m_e c^2$	0.511004 MeV
Charge-to-mass ratio, e/m_e	1.7588×10^{11} C/kg

THE PHOTOELECTRIC EFFECT

At about the time that Thomson was isolating and studying electrons, it was observed that when ultraviolet (UV) light irradiates a clean zinc plate, the plate acquires a positive electric charge (Fig. 10.1). The radiation literally knocks electrons off the plate, leaving behind a positive charge. This phenomenon of electron ejection by light is called the *photoelectric effect*.

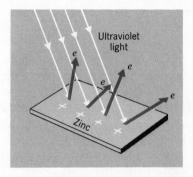

Fig. 10.1 *The photoelectric effect. Ultraviolet light incident on a metal plate, such as zinc, ejects electrons and the plate acquires a positive charge.*

The qualitative explanation of the photoelectric effect in terms of the removal of electrons by the action of light was in no way revolutionary. Light was known to consist of oscillating electric and magnetic fields that carry electromagnetic energy. It was entirely consistent with classical theory that these waves could transfer energy to electrons in the metal and when an electron had acquired sufficient energy from the wave it could escape from the metal. But further experiments showed that several aspects of the photoelectric effect were at variance with the predictions of classical electromagnetic theory.

Quantitative information regarding the photoelectric effect was obtained by using apparatus similar to that shown schematically in Fig. 10.2. Two plates, *A* and *B*, are contained within an evacuated tube and are connected to a source of variable voltage and a sensitive current-measuring instrument (a galvanometer or an ammeter). Plate *A* is the photoemissive surface and electrons are ejected from this plate when it is exposed to ultraviolet radia-

THE FOUNDA-TIONS OF QUANTUM THEORY

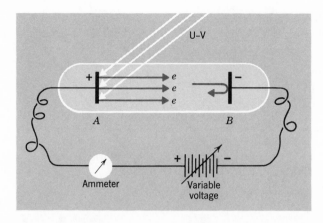

Fig. 10.2 *Apparatus for the study of the photoelectric effect. The motion of the energetic electrons emitted from plate* A *is retarded by the negative potential of plate* B *relative to* A.

tion; plate *B* is the collector plate. With this apparatus the current of photoelectrons can be measured as a function of the frequency and the intensity of the UV radiation and of the voltage between plates *A* and *B*. Also, the effect of using different materials for the photoemissive surface can be studied.

One of the important measurements that can be made in the investigation of the photoelectric effect is the determination of the maximum retarding (or stopping) potential V_r that can be overcome by the energetic photoelectrons ejected when plate *A* is irradiated with UV light of a definite frequency. By making plate *B* negative with respect to plate *A*, the electrons are slowed down as they move from *A* to *B*. If the retarding potential is sufficiently great, no electrons can reach plate *B* and the ammeter will register no current flow. Therefore, by increasing the negative potential of *B* and observing the current flow, the voltage V_r necessary to just stop the electrons from reaching plate *B* can be determined. If an electron with kinetic energy *KE* is just stopped when the potential difference between the plates is V_r then $KE = eV_r$. That is, eV_r is the *maximum* kinetic energy of the ejected photoelectrons. (Electrons liberated below the surface of the material can lose kinetic energy before emerging from the surface.) Measurements of this type led to the important result that V_r (and, hence, $KE_{\max}$) increases in direct proportion to the *frequency* of the incident UV radiation. Furthermore, there was found to be a *threshold frequency* v_o below which no photoelectrons could be produced *regardless of the intensity of the incident radiation*. Each material that is used for the photoemissive surface has its own characteristic threshold frequency (see Fig. 10.3). Classical electromagnetic theory was powerless to explain these results.

EINSTEIN'S EXPLANATION OF THE PHOTOELECTRIC EFFECT

In 1905 Einstein offered a theory that provides an explanation of the entire set of facts concerning the photoelectric effect. Einstein's photoelectric theory was beautifully simple, with the same brevity and elegance that characterized his relativity theory (proposed in the same year). Einstein drew upon a proposal that had been made a few years earlier by the noted German physicist Max Planck (1858–1947) to the effect that energy exchanges take place in discrete steps by *quanta*. Planck's unorthodox hypothesis was not

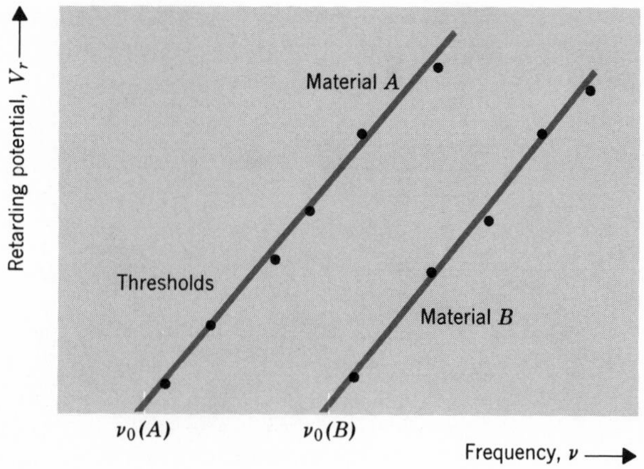

Fig. 10.3 *The maximum retarding potential that can be overcome by photoelectrons is directly proportional to the UV frequency for all frequencies above the threshold at $v = v_0$. The dots show the results of various measurements made with discrete UV frequencies for two different materials. The two lines have the same slope but different thresholds.*

taken seriously at the time because the traditional view was that energy could be exchanged in arbitrary amounts; that is, energy was thought to be a *continuous* quantity, just as length and time were (and still are) considered to be continuous. Einstein extended Planck's idea and proposed that *all* electromagnetic radiation exists in the form of discrete quanta or *photons*. The amount of energy associated with a quantum of frequency v is a certain constant times v; that is,

$$\mathcal{E} = hv \tag{10.1}$$

where h is the proportionality constant, now known as *Planck's constant*. The value of the constant h is extremely small:

$$h = 6.625 \times 10^{-34} \text{ J-sec} \tag{10.2}$$

Therefore, each quantum has associated with it an energy that is very small by ordinary standards.

Einstein further proposed that when a photon interacts with matter it behaves in almost the same way as a *particle* and delivers its energy, not to the material as a whole or even to an atom as a whole, but to an *individual* electron. The occurrence of a threshold is due to the fact that a certain amount of energy must be supplied to the electron in order to free it from the material (even if it receives no kinetic energy). Furthermore, different materials have different values for the threshold energy.

According to Einstein, then, the kinetic energy of a photoelectron must be the difference between the energy of the incident UV photon and the minimum energy necessary to free the electron from the material (called the *work function* of the material). That is,

$$\text{(electron } KE) = \text{(photon energy)} - \text{(work function)} \tag{10.3}$$

When hv is used for the photon energy, this equation becomes

$$KE = hv - \phi \tag{10.4}$$

where ϕ is the work function for the particular material.

THE FOUNDA-
TIONS OF
QUANTUM
THEORY

As a mechanical analog to this explanation of the photoelectric effect, consider a ball of mass m that is at rest in a trough, as in Fig. 10.4. If a sufficient amount of energy $\mathcal{E}$ is supplied to the ball, it will roll up the side of the trough of height H and "escape" with a final velocity v. The energy equation for the process is

$\frac{1}{2}mv^2 = \mathcal{E} - mgH$

In this expression, mgH is the "work function" (that is, the potential energy barrier that must be overcome) and $\mathcal{E}$ is equivalent to the photon energy.

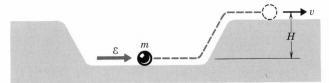

Fig. 10.4 *A mechanical analog of the photoelectric effect. Sufficient energy must be supplied to the ball for it to "escape" from the trough and proceed away with a velocity* v.

Example **10.1**

What will be the maximum kinetic energy of the photoelectrons ejected from magnesium (for which $\phi = 3.7$ eV) when irradiated by UV light of frequency 1.5×10^{15} sec^{-1}?

The energy of a photon with frequency 1.5×10^{15} sec^{-1} is

$h\nu = (6.6 \times 10^{-34} \text{ J-sec}) \times (1.5 \times 10^{15} \text{ sec}^{-1})$

$\quad = (9.9 \times 10^{-19} \text{ J}) \times \left(\dfrac{1}{1.6 \times 10^{-19} \text{ J/eV}}\right)$

$\quad = 6.2$ eV

Therefore, the maximum kinetic energy is

$KE = h\nu - \phi = 6.2 \text{ eV} - 3.7 \text{ eV} = 2.5 \text{ eV}$

Einstein's explanation of the photoelectric effect was the first convincing demonstration of the quantum nature of electromagnetic radiation. For this work (*not* for his relativity theory!), Einstein was awarded the Nobel Prize in 1921.

10.2   *Waves and/or Particles?*

THE WAVE-PARTICLE DILEMMA

The wave character of light was established in the early years of the 19th century when a series of interference and diffraction experiments disproved the competing particle theory of light. But Einstein's photoelectric theory revived the notion that light behaves as a particle—at least, in its interactions with atomic electrons. Does this mean that we are forced to abandon the wave theory and return to the old particle theory? Or is there some peculiar

Table **10.2** *Photoelectric Properties of Some Elements*

Element	Work Function, ϕ (eV)	Threshold Frequency (sec⁻¹)	Threshold Wavelength (Å)	
Cesium	1.9	4.6×10^{14}	6500	Visible light
Potassium	2.2	5.3	5600	
Sodium	2.3	5.6	5400	
Calcium	2.7	6.5	4600	
Magnesium	3.7	8.9	3400	Ultraviolet
Silver	4.7	11.4	2600	
Nickel	5.0	12.1	2500	

feature of light that presents a two-sided appearance, sometimes wave-like and sometimes particle-like? If so, then how do we know when to expect one feature and not the other? These were the questions about light that were raised by the proposal of the quantum nature of electromagnetic radiation.

THE DE BROGLIE WAVELENGTH

Einstein's explanation of the photoelectric effect clearly demonstrated that electromagnetic radiation possess properties that closely resemble those of material particles. In 1924 a young Frenchman, Louis Victor de Broglie (1892–   ), proposed a complementary hypothesis, namely, that particles can exhibit *wave*-like properties.

In order to describe a wave, there must be a definable wavelength for the propagation. Consider a photon that has an energy $\mathcal{E}$. According to the Einstein mass-energy relation (Eq. 5.24), a mass m has a mass-energy $\mathcal{E} = mc^2$. Or, conversely, an entity that possesses a total energy $\mathcal{E}$ has associated with it an equivalent mass $m = \mathcal{E}/c^2$. And since electromagnetic radiation travels with a velocity c, the momentum of the radiation (that is, equivalent mass $\times$ velocity) must be

$$p = m \times c = \frac{\mathcal{E}}{c^2} \times c = \frac{\mathcal{E}}{c}$$

Then, using $\mathcal{E} = h\nu$ and $\nu/c = \lambda$, we can express the momentum as

$$p = \frac{\mathcal{E}}{c} = \frac{h\nu}{c} = \frac{h}{\lambda}, \quad \text{or} \quad \lambda = \frac{h}{p}$$

De Broglie argued that the wavelength of a material *particle* should follow exactly the same prescription:

de Broglie wavelength: $\boxed{\lambda = \dfrac{h}{p}}$ (10.5)

THE FOUNDA-TIONS OF QUANTUM THEORY

Within three years after de Broglie's ingenious proposal (which earned him the 1929 Nobel Prize), the wave properties of electrons had been demonstrated in experiments by Davisson and Germer in America and by

G. P. Thomson in England which showed the diffraction of electrons. These experiments are described briefly in the following paragraph.

Example **10.2**

What is the wavelength of a 10-eV electron?

For an electron of this low energy we can use the nonrelativistic expression without appreciable error. Since $KE = \frac{1}{2}mv^2$ and $p = mv$, we have

$$p = \sqrt{2m_e KE}$$

so that

$$\lambda = \frac{h}{p} = \frac{h}{\sqrt{2m_e KE}}$$

$$= \frac{6.62 \times 10^{-34} \text{ J-sec}}{\sqrt{2 \times (9.11 \times 10^{-31} \text{ kg}) \times (10 \text{ eV}) \times (1.60 \times 10^{-19} \text{ J/eV})}}$$

$$= 3.86 \times 10^{-10} \text{ m} = 3.86 \text{ Å}$$

A *photon* with this same wavelength would have an energy

$$\mathcal{E} = h\nu = \frac{hc}{\lambda}$$

$$= \frac{(6.62 \times 10^{-34} \text{ J-sec}) \times (3 \times 10^8 \text{ m/sec})}{(3.86 \times 10^{-10} \text{ m}) \times (1.60 \times 10^{-19} \text{ J/eV})} = 3.2 \text{ keV}$$

Such an energetic photon is an *X ray*.

DIFFRACTION OF X RAYS AND ELECTRONS

When electrons are allowed to fall through a potential of several thousand volts and strike a metal target, extremely short-wavelength radiation is produced. Many attempts were made in the early 1900s to measure the wavelengths of these X rays by conventional diffraction experiments. These experiments were only marginally successful until Max von Laue (1879–1960) realized that the regular planes of atoms in crystals, with spacings of only a few Ångstroms, could be used to diffract X rays. In 1912 von Laue succeeded in obtaining interference patterns of X rays diffracted from calcite crystals. The spacing between the planes of atoms in the crystal could be calculated approximately from a knowledge of the properties of the material. The measured diffraction patterns then showed that typical wavelengths for X rays were of the order of 1 Å, very much shorter than UV wavelengths.

The diffraction of X rays is illustrated schematically in Fig. 10.5 where the dots represent the ordered array of atoms in a simple crystalline lattice. The rows of atoms are spaced a distance d apart and therefore form a kind of grating (or series of slits) through which the radiation can pass. In the direction specified by the angle θ with respect to the direction of the incident beam, the two scattered rays differ in their path lengths by an amount Δ. Therefore, on a photographic plate placed some distance away, the interference between the two rays will be *constructive* if Δ is an integer number of wavelengths of the radiation; the interference will be *destructive* if Δ is

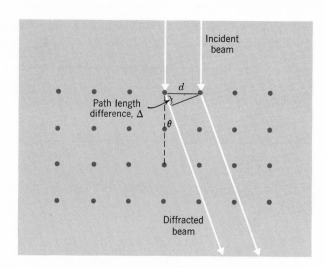

Fig. 10.5 *An incident beam of X rays (or electrons) is diffracted by the grating formed by the regular array of atoms in a crystal.*

an odd number of half wavelengths. Consequently, there will be exposed and unexposed regions that alternate on the photographic plate. A *two*-dimensional array such as this produces a pattern of *lines* (see Fig. 8.14). But a crystal is a *three*-dimensional array of atoms. Consequently, the directions in space in which the interference is constructive are severely restricted and the exposed regions on the photographic plate are *spots* instead of lines. Figure 10.6*a* shows such a pattern of spots (called a *Laue* pattern) produced by X rays incident on a crystal of rock salt (NaCl).

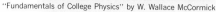

"Fundamentals of College Physics" by W. Wallace McCormick

Wollan, Shull, & Marney, Physics Rev. 73, 527 (1948)

Fig. 10.6 *Laue patterns showing the diffraction of (a) X rays and (b) neutrons by single crystals of rock salt. The light regions are the exposed regions of the photographic plates caused by constructive interference. Note the similarity of the interference patterns. The central bright spot in each pattern is due to the undeflected portion of the beam.*

THE FOUNDA-
TIONS OF
QUANTUM
THEORY

According to de Broglie's hypothesis, electrons and other material particles should exhibit wave properties and therefore should be capable of producing diffraction patterns in the same way that X rays produce these patterns. In 1927 George P. Thomson (1892–   , son of Sir J. J. Thomson) observed electron diffraction by passing electrons through foils. At almost the same time, C. J. Davisson (1881–1958) and L. H. Germer (1896–   ) directed electrons onto the surfaces of single crystals and found interference peaks in the distribution of the reflected electrons. Davisson and Thomson shared the 1937 Nobel Prize for these experiments that established the wave character of electrons.

The correspondence between X-ray diffraction and particle diffraction is strikingly illustrated in Fig. 10.6. The photograph on the right shows the diffraction pattern for neutrons incident on a crystal of rock salt. This pattern is almost identical to the pattern produced by X rays on the same material (Fig. 10.6a). The close similarity between the patterns in this pair of photographs is dramatic proof of the identical wave properties of electromagnetic radiation and matter.

10.3   *The Basis of Quantum Theory*

THE INTERFERENCE OF ELECTRON WAVES

Let us consider a very simple type of experiment involving electrons. Suppose that we direct a beam of electrons toward a panel in which two slits have been cut. In Fig. 10.7a we show the situation for the case in which the lower slit (labeled *B*) is blocked off so that all of the electrons that penetrate the panel must go through slit *A*. At a certain distance behind the panel we place a screen for the purpose of observing the positions of impact of the various individual electrons. Such a screen can be made from a material that *scintillates;* that is, a flash of light is produced for every electron impact. Using this technique we can measure the distribution of electron impacts along the screen. The curve in Fig. 10.7a represents such an *intensity* distribution; it shows that the peak of the distribution is directly in line with the slit. If the slit is *narrow,* the distribution will be *broad* (compare Fig. 8.17).

If we repeat the experiment with slit *B* open and slit *A* closed, we shall, of course, find exactly the same type of distribution centered now around the position in line with Slit *B*.

What will happen if we open *both* slits? If electrons behaved as little balls, we could think of the experiment in terms of projecting a stream of marbles at the pair of slits; some marbles would go through slit *A* and, because of scattering at the slit edges, would produce a pattern on the screen like that shown in Fig. 10.7a and some would go through slit *B* and produce the pattern of Fig. 10.7b. The net result would be the *sum* of the two intensity patterns. But this expectation based on classical reasoning is not borne out by experiment for the case of electron beams. Actually, a much more complex pattern is observed (Fig. 10.7c). The maximum intensity is found *midway* between the two slits and there are several subsidiary maxima, falling off in intensity uniformly on each side of the central maximum. Notice also that, at some points on the screen, the intensity has actually been *decreased* by the opening of an additional slit. We can only draw the conclusion that electrons behave as waves in this experiment and produce exactly the same type of interference effects that are produced by light waves.

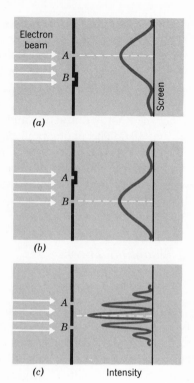

Fig. 10.7 (a, b) *Intensity patterns of electron impacts on a screen for a single open slit.* (c) *Intensity pattern for a double slit. This pattern is not the same as the sum of the two previous patterns. This pattern is drawn for the case in which the distance between the two slits is 3 times the width of each slit. Notice that increasing intensity is plotted to the left.*

SELF-INTERFERENCE AND PROBABILITY

Suppose that we repeat the double-slit experiment with electrons (we could equally well use photons), but now we make the incident beam exceedingly weak—so weak, in fact, that at any one instant there can be only a single electron in the vicinity of the apparatus. What pattern of light flashes on the screen will result? What is the relationship between the intensity pattern found for a *beam* of electrons and the way in which an *individual* electron will behave in the apparatus?

We have already established that electrons exhibit the wavelike property of diffraction when they pass through slits. But we also know that a single electron will strike the screen only at one point. (We see in this experiment the appearance of both the wavelike and the particlelike properties of the electron during different phases of the experiment.) Where will the first electron strike the screen? We do not know! But on the basis of the previous experiment with a beam of electrons we can anticipate that the electron will *probably* strike the screen at the point corresponding to the peak of the intensity curve (see Fig. 10.7c). This expectation is, in fact, correct; the electron *could* strike the screen at *any* point but the probability is highest that it will strike at the position of the intensity maximum. For the second electron the probability of striking the screen at any particular position is exactly the same as for the first electron. Some of the electrons will strike at the intensity maximum and others will strike at different positions. After the first few electrons have passed through the apparatus, a graph of the observed light flashes will show only a scatter of points. However, when thousands of electron impacts have been observed, we find that the distribution is exactly the same as the intensity curve found for a beam of electrons. We are therefore led to the conclusion that the *intensity* pattern for the beam of electrons corresponds precisely to the *probability* curve that specifies the liklihood that any single electron will strike the screen at a particular position.

Individual electrons (or photons) exhibit wave properties, complete with the phenomenon of *interference*. When a single electron is in the apparatus at a given instant, it has only *itself* with which to interfere. This self-interference results from that portion of the electron wave that goes through slit *A* interfering with the portion that goes through slit *B*. Only when large numbers of particles are involved does it become possible to identify the intensity curve with the diffraction pattern predicted by the wave theory.

IS THE WAVE-PARTICLE DUALITY REAL?

How can we reconcile the fact that both electrons and photons appear sometimes as particles and sometimes as waves? Does each exist as part wave and part particle? Or are they both capable of transforming back and forth between these two different descriptions? The answers to these questions become clear by realizing that when we make a wave or a particle classi-fication we have forced a *classical* or Newtonian description on entities that are essentially *nonclassical* and do not obey the rules of Newtonian me-chanics. Electrons and photons do not obey the rules of classical mechan-ics—their behavior is described correctly only by *quantum* mechanics. It is therefore not surprising that certain ambiguities arise when we use classical ideas to describe quantum objects.

In the quantum mechanical view of Nature, an experimenter's apparatus and the object under study together constitute a system. The only meaningful way to discuss the behavior of the object studied is in terms of the results of *measurements*. Therefore, whether an electron or a photon appears as a wave or as a particle depends on the nature of the measurement that is made. The wavelike or particlelike character of an electron or a photon therefore lies only in the eye of the beholder.

10.4   *The Quantization of Momentum and Energy*

THE FREE PARTICLE

A quantum mechanical *free* particle can move through space without any restriction on the value of the energy that it can have. Thus, the particle wave can have any wavelength $\lambda = h/p$, and it can have any kinetic energy, which, in a nonrelativistic situation, is expressed as

$$KE = \frac{1}{2}mv^2 = \frac{1}{2}\frac{(mv)^2}{m} = \frac{p^2}{2m} \tag{10.6}$$

The relationship between the kinetic energy and the momentum, therefore, is *parabolic*, as shown in Fig. 10.8, every point on the curve represents an allowed energy and the corresponding allowed momentum.

For the allowed energies of a free particle, there is no difference between the results of classical mechanics and quantum mechanics—all energies are

$$KE \quad KE = \frac{p^2}{2m}$$

Momentum, $p \longrightarrow$

Fig. 10.8 *For a free particle, the curve that relates the kinetic energy to the momentum is a parabola, and every point on the curve refers to an allowed energy and the corresponding allowed momentum.*

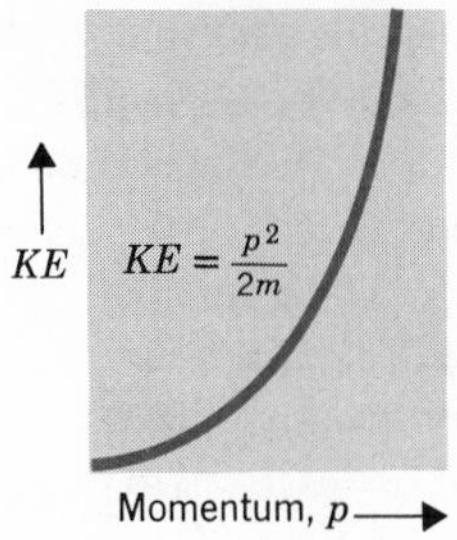

allowed. However, if we *confine* the particle in some way by subjecting it to forces that restrict its motion, the two theories no longer yield the same results.

PARTICLE BOUND IN A ONE-DIMENSIONAL "BOX"

In many situations we have to deal, not with *free* particles, but with particles that are constrained in some way to remain in a certain region of space. A simple example of this situation is that in which a particle is required to move along a straight line (for example, the *x*-axis) between the points $x = 0$ and $x = L$. Classically, we can think of this particle as bouncing between a pair of unyielding walls, always maintaining straight-line motion in the $+x$- or $-x$-direction. From the standpoint of classical theory, there is again no restriction on the energy that the particle can have. Energy and momentum are still related by Eq. 10.6 and any combination of *KE* and *p* that satisfies this condition is allowed.

Now consider a quantum particle (for example, an electron) that is required to move in the same way. We must think of *this* particle in terms of its wave character and, in particular, we must examine the conditions imposed on the electron wave by the presence of the walls. It proves convenient to describe the electron in terms of a *wave function,* denoted by the symbol ψ. The value of the wave function depends on position and in the one-dimensional case is written as $\psi(x)$. The square of the wave function, $|\psi(x)|^2$, is proportional to the probability that a measurement will reveal the electron to be located at the particular position specified by *x*. In fact, the wave function itself, $\psi(x)$, has no physical significance and cannot be directly measured; only $|\psi(x)|^2$ can be determined by experiment and is therefore physically meaningful.[1]

If the particle is confined to a "box," then there is zero probability of finding the particle outside the box, and hence, the wave function must be zero in this region. Because the wave function cannot change abruptly from some nonzero value just inside the box to zero just outside the box, the wave function must also be zero exactly at the boundary walls. With these facts in mind, the solution to the problem is now quite simple because it is exactly the same as the problem of standing waves on a string (Section 8.2). The wave function for the particle (just as the displacement or ampli-

THE FOUNDA-
TIONS OF
QUANTUM
THEORY

[1] In classical physics the wave amplitude *does* have physical meaning and can be measured. In both classical and quantum physics the square of the wave amplitude is proportional to the *intensity* of the wave.

tude of the string) must be zero at $x = 0$ and $x = L$. That is, standing de Broglie waves must be fitted into the box, and this can be accomplished only when an integer number of wavelengths is equal to $2L$. Thus,

$$n\lambda_n = 2L, \qquad n = 1, 2, 3, \ldots \tag{10.7}$$

The first 4 allowed ψ-waves are illustrated in Fig. 10.9.

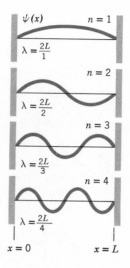

Fig. 10.9 *For a quantum particle bound in a one-dimensional "box," standing de Broglie waves (ψ-waves) must be fitted into the "box." The allowed wavelengths are therefore integer fractions of $2L$.*

The probability of finding the particle as a particular point within the box is proportional to $|\psi(x)|^2$. This quantity is shown in Fig. 10-10 for the case $n = 4$. Notice that there are 4 regions in which there is a high probability of finding the particle and that there is zero probability not only at the walls but also at certain points *within* the box. Clearly, this result is contrary to that of classical mechanics.

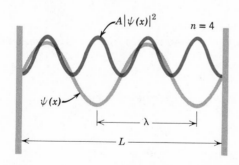

Fig. 10.10 *The probability amplitude $\psi(x)$ and the intensity $A|\psi(x)|^2$ for the case $n = 4$. The particle is most likely to be found at discrete positions within the "box."*

ALLOWED ENERGIES IN THE "BOX"

We can now calculate the energies corresponding to the allowed wavelengths by using the de Broglie relation and Eq. 10.7. The allowed momenta are

$$p_n = \frac{h}{\lambda_n} = n\frac{h}{2L}, \qquad n = 1, 2, 3, \ldots \tag{10.8}$$

and the corresponding energies are

$$KE_n = \frac{p_n^2}{2m} = n^2 \left(\frac{h^2}{8mL^2} \right), \qquad n = 1, 2, 3, \ldots \qquad (10.9)$$

Therefore, we have the important result that only certain discrete energies and momenta are allowed; energy and momentum are *quantized*. Instead of the classical result in which every point on the *KE* versus *p* parabola corresponds to a possible energy-momentum combination, the quantum result shows that only certain of these points are in fact allowed (see Fig. 10.11).

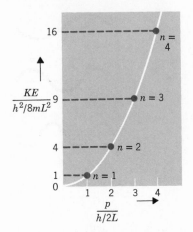

Fig. 10.11 *Only discrete energies (indicated by the horizontal lines and the dots) are allowed for a quantum particle in a box.*

Another important point to realize is the fact that *zero* kinetic energy for the particle is *not* allowed—the particle cannot be at rest in the box. The state of rest requires zero momentum and, hence, an infinitely long wavelength. Such a de Broglie wave cannot fit into any finite box and therefore is not allowed. In general, no quantum system (with the exception of the ideal free particle) can possess zero kinetic energy. Even at the absolute zero of temperature, when according to classical theory all motion must cease, a quantum system still possesses a certain kinetic energy, called the *zero-point energy*.

10.5 *The Uncertainty Principle*

WHERE DO THE ELECTRONS "REALLY" GO?

When a series of individual electrons, spaced in time, are incident on a double slit, it is necessary to treat each electron as if its "probability wave," described by $\psi(x)$, goes through *both* slits. The interference of the two portions of each wave at the screen determines the distribution of light flashes that is observed. But an electron is a *particle* and is indivisible; no one has ever observed a *part* of the mass or a *part* of the charge of an electron. Our intuition tells us that an electron cannot go through two separated slits—it must go through one or the other. Can we perform an experiment to determine where the electron "really" goes? Let us alter the double-slit experiment by placing a thin detector behind one of the slits so that the electrons will still be able to pass through but will give a signal when this occurs. Then, we allow electrons to enter the apparatus one at a time and we record only those flashes of light on the screen that are accompanied by a signal that indicates the electron went through the slit with the detector. What pattern of flashes do we find? We find exactly the *single*-slit pattern again (Fig. 10.7a)! The act of determining which slit the electron went through has destroyed the double-slit interference effect.

Perhaps the trouble with our experiment was the detector. Perhaps it was not sufficiently "thin" and actually disrupted too severely the electron trajectories passing through it. We can repeat the experiment using a source of light behind one of the slits. Whenever we detect a photon that is scattered by an electron passing through the slit, we arrange to record the position of impact of the electron on the screen. But this technique is no more successful than the first; the results of this experiment are the same as before. In fact, whatever method we devise to indicate that the electrons have gone through a particular slit, the result is always that the interference effect is destroyed.

So our intuition is wrong. Our insistence in thinking of electrons in terms of classical particles leads to inconsistencies. Because of the difficulties that arose in applying classical reasoning to individual events in the atomic domain, the German theorist Werner Heisenberg (1901–) concluded that there must be a general principle of Nature that places a limitation on the capabilities of all experiments. This principle, formulated in 1927, is known as the *uncertainty principle*.

According to Heisenberg's principle, it is impossible to build a detector to determine through which slit the electron passed without destroying the interference pattern. That is, there can be no device that can reveal the presence of an electron with a sufficiently delicate touch that the interference pattern will be unaffected—the act of "looking at" an electron with even a *single* photon is sufficient to change the wave function of the electron and disrupt the interference pattern.

STATEMENT OF THE UNCERTAINTY PRINCIPLE

The concept of a *particle* is something localized in space. According to classical theory a particle has, at a given instant, both a well-defined position and a well-defined velocity. Let us attempt to apply this reasoning to an elementary particle, such as an electron.

What is the restriction, if any, on determining the location of an electron with an imaginary super-high-power microscope? Because we use light (or higher energy photons, such as X rays) to determine position with a microscope, we must remember the wave nature of such radiation. In fact, we cannot locate the electron in this way to better than about one wavelength of the radiation; that is, the uncertainty in the position Δx is approximately equal to one wavelength: $\Delta x \approx \lambda$. But in the act of observing the position of the electron, at least one photon must have been scattered by the electron. Thus, the uncertainty in the momentum of the electron will be equal to the momentum imparted to it by the photon, which (using the de Broglie relation) will be $\Delta p_x \approx h/\lambda$. The product of Δx and Δp_x is, therefore, $\Delta x\, \Delta p_x \approx \lambda \times (h/\lambda)$. Thus,

$$\Delta x\, \Delta p_x \approx h \tag{10.10}$$

There are many other methods for determining this expression for Heisenberg's uncertainty principle and there are many experiments (including *thought* experiments) that can be devised to test its validity. *No* experiment has ever been found to be in disagreement with the uncertainty principle.

It is important to realize that the uncertainty principle refers to the *predictability* of events. *After* an electron goes through the slit in Fig. 10.7*a* and strikes the screen, we know where it did so from the position of the light flash, but before the event takes place, we can only give the *probability* that a light flash will be observed at a particular point. Quantum theory cannot predict the result of any single event, but the *average* of a large number of events (for example, the position of the peak of the probability or intensity curve) can be predicted with precision. This is the essential meaning of the uncertainty principle.

The uncertainty principle is not to be thought of as some mysterious device conceived by Nature to prevent man from probing too deeply into her methods of making atoms behave properly. Rather, the uncertainty principle is just one manifestation of the wave-particle duality of radiation and matter. Waves cannot be localized in space and so any measurement of the position of a wavelike object must be subject to uncertainty. The Heisenberg principle gives a quantitative description of this uncertainty.

Example **10.3**

The position of a free electron is determined by some optical means to within an uncertainty of 10,000 Å or 10^{-6} m. What is the uncertainty in its velocity? After 10 sec how well will we know its position?

Nonrelativistically, we can express the uncertainty relation as

$$\Delta p_x = m_e \times \Delta v_x \approx \frac{h}{\Delta x}$$

so that

$$\Delta v_x \approx \frac{h}{m_e\, \Delta x} = \frac{6.6 \times 10^{-34}\ \text{J-sec}}{(9.1 \times 10^{-31}\ \text{kg}) \times (10^{-6}\ \text{m})} \cong 700\ \text{m/sec}$$

Therefore, after 10 sec the electron could be anywhere within a distance of 7×10^3 m or 7 km! The act of locating the electron at one instant to within a distance as small as 10^{-6} m stringently limits our knowledge of where the electron is at future times.

ANOTHER FORM OF THE UNCERTAINTY PRINCIPLE

In 1928, Niels Bohr summarized the conclusions that had been reached concerning indeterminism in quantum theory by stating that *if an experiment allows us to observe one aspect of a physical phenomenon, it simultaneously prevents us from observing a complementary aspect of the phenomenon.* This statement is known as Bohr's *principle of complementarity*. The complementary features to which the principle applies may be the position and momentum of a particle, or the wave and particle character of matter or radiation. In fact, any pair of a particle's dynamical variables that have the same product of dimensions obey a similar uncertainty relation. The dimensions of h are J-sec (the dimensions of *angular momentum*); the product *energy* $\times$ *time* has the same dimensions and the corresponding uncertainty relation is

$$\boxed{\Delta \mathcal{E}\, \Delta t \approx h}$$ (10.11)

Example **10.4**

In emitting a photon, an atom radiates for approximately 10^{-9} sec. What is the uncertainty in the energy of the photon?

$$\Delta \mathcal{E} \approx \frac{h}{\Delta t} \cong \frac{6.6 \times 10^{-34}\ \text{J-sec}}{10^{-9}\ \text{sec}} \times \frac{1}{1.6 \times 10^{-19}\ \text{J/eV}} \cong 4 \times 10^{-6}\ \text{eV}$$

Summary of Important Ideas

The photoelectric effect can be explained only if electromagnetic radiation occurs in discrete packets or *photons*.

The fundamental atomic constants are: the velocity of light c, the mass of the electron m_e, the charge of the electron e, and Planck's constant h.

Depending on the type of measurement that is made, electrons and photons can exhibit properties of either *waves* or *particles*.

A particle has associated with it a *de Broglie wavelength* $\lambda = h/p$. Radiation has associated with it an equivalent mass $m = \mathcal{E}/c^2$, and a momentum $p = \mathcal{E}/c$.

In a double-slit experiment, *no* measurement can be made to determine through which slit the photon or electron went without destroying the double-slit interference pattern.

Because of the wave nature of radiation and particles, we can never predict the exact behavior of any particular photon or particle; we can only predict the *average* behavior of large numbers of photons or particles. Individual events can be discussed only in terms of *probabilities*.

A particle or a photon is described in terms of a quantum mechanical wave function $\psi(x)$. Only the *square* of the wave function, which is proportional to the *intensity*, can be measured and therefore has physical meaning.

Except for the (ideal) free particle, all quantum mechanical systems are constrained to have only certain discrete energies and momenta.

The *Heisenberg uncertainty principle* expresses the fact that we cannot simultaneously measure with arbitrarily high precision *complementary* aspects of a particle or a photon (such as *momentum* and *position* or the *energy* of an event and the *time* interval during which it took place).

Questions

10.1

Herman von Helmholtz once said that he was puzzled to explain what an electric charge (that is, an electron) is, except the recipient of a symbol. Comment on the view that an *electron* is just a name that is convenient for describing various observations rather than a *thing*.

10.2

Describe an experiment to distinguish between an X ray whose wavelength is $\lambda_X = 10^{-10}$ m and an electron whose de Broglie wavelength is $\lambda_e = 10^{-10}$ m. What experiments would *not* be suitable?

10.3

Discuss some of the changes in everyday events that would result if Planck's constant were suddenly increased to 1 J-sec.

10.4

Refer to Fig. 10.10 where it is shown that the probability of finding a particle at a given position in a "box" is a maximum at certain positions and is *zero* at other positions. How is it possible for the particle to "move" from one position of maximum probability to another since in order to do so it must pass through a position that it is not allowed to occupy? Explain the situation carefully in terms of measurements that can be made.

10.5

According to John A. Wheeler, the following two items are complementary in the sense of Bohr's principle of complementarity: (a) the use of a word to convey *information*, (b) the analysis of the *meaning* of the word. Discuss complementarity in this case.

10.6

One of the reasons cited to justify the construction of expensive high-energy accelerators is that particles (and photons) of high energy are needed to probe the detailed structure of nuclei and nucleons. Why is this so?

10.7

Choose a quantity that was once thought to be continuous and discuss how it was shown to be discrete. Choose a quantity that was once thought to be discrete and discuss how it was shown to be "fuzzy."

10.8

Discuss the proposition that we shall eventually be able to overcome the limitations now set by the uncertainty principle and shall then be able to discuss microscopic phenomena with the same kind of deterministic approach that is appropriate for macroscopic mechanical phenomena (that is, the approach of Newtonian dynamics).

THE FOUNDA-
TIONS OF
QUANTUM
THEORY

230

Problems

10.1

The threshold wavelength for the photoelectric effect on a certain material is 3000 Å. What is the work function of the material?

10.2

A potential of 2.7 volts is required to stop completely the photoelectrons from a certain material when the electrons are ejected by 2100-Å radiation. What is the work function of the material?

10.3

The work function for platinum is 5.32 eV. What is the longest wavelength photon that can eject a photoelectron from platinum?

10.4

The work function of barium is 2.48 eV. What is the maximum kinetic energy of a photoelectron ejected from barium by photons of wavelength 2000 Å?

10.5

Under certain conditions the retina of the human eye can detect as few as five photons of blue-green light ($\lambda = 5 \times 10^{-7}$ m). What is the corresponding amount of energy received by the retina in joules and in eV?

10.6

What is the mass equivalent of a photon of wavelength 6×10^{-7} m (yellow light)? How many such photons would be required to make up the rest-mass energy of one electron?

10.7

What is the frequency of radiation whose wavelength is 1 Å? What energy is carried by such a photon?

10.8

Through what potential difference must an electron fall (starting at rest) in order that its wavelength be 1.6 Å?

10.9

A proton, an electron, and a photon all have de Broglie wavelengths of 1 Å. If they all leave a given point a $t = 0$, what are the arrival times at a point 10 m away?

10.10

What is the velocity of a helium atom that has a wavelength of 1 Å?

10.11

When a neutron is in thermal equilibrium with objects at room temperature it has an energy of 0.025 eV; such neutrons are called *thermal* neutrons. What is the wavelength of a thermal neutron?

10.12

What is the energy of a photon whose wavelength is (a) the size of an atom (10^{-10} m), (b) the size of a nucleus (5×10^{-15} m)?

10.13

An electron is localized in the x-direction to within 1 mm. How precisely can its x-momentum be known?

10.14

A proton in a nucleus is localized to within a distance approximately equal to the nuclear radius. What is the approximate uncertainty in the velocity of a proton in an iron nucleus ($R \cong 6 \times 10^{-15}$ m)? What is the corresponding uncertainty in energy? (A nonrelativistic calculation is adequate for the accuracy desired.)

10.15

The electron in a hydrogen atom may be considered to be confined to a region of radius 5×10^{-10} m around the nucleus. Use the uncertainty principle to estimate the momentum of the electron and, from this, its kinetic energy. (For purposes of estimating these quantities, make only a one-dimensional calculation; that is, use $\Delta x \, \Delta p_x \simeq h$.) Why does the electron remain attached to the nucleus?

10.16

A certain atomic energy level has a mean lifetime of 10^{-8} sec. (That is, after the level is excited by some means, it requires, on the average, a time of 10^{-8} sec before it spontaneously radiates a photon.) What is the uncertainty in the energy of the emitted radiation?

10.17

The photons emitted by atoms excited to a certain level are found to vary in energy from 10.00001 eV to 10.00003 eV. What is the average lifetime of the level?

THE
FOUNDA-
TIONS
OF
QUANTUM
THEORY

232

CHAPTER 11
ATOMS AND MOLECULES

The sciences of astronomy and chemistry played crucial roles in establishing the foundations of physics. Early astronomical observations and measurements provided the basis for the theory of gravitation (Chapter 4)—the first of the great physical theories and the key to the understanding of large-scale phenomena in the Universe. Similarly, 19th century chemistry established the atomistic character of matter and opened the door for physicists to develop a *microscopic* description of Nature.

Thomson discovered the electron in the closing decade of the 19th century, but the source of positive charge in atoms remained a mystery for another 15 years until Rutherford provided the explanation in terms of his nuclear model of the atom. Niels Bohr was quick to adopt this new idea and his early work on the details of atomic structure set the stage for the tremendous outpouring of theoretical and experimental results that, in the short space of four years from 1924 to 1928, firmly established quantum theory as a proper description of atomic processes.

11.1   *The Nuclear Atom*

RUTHERFORD'S SCATTERING EXPERIMENT

At the time that Thomson was studying the properties of electrons, Ernest Rutherford (1871–1937) (Fig. 11.1) was investigating the newly discovered α *rays,* positively-charged particles that are emitted with high velocities in certain types of radioactive decay processes. By 1908, Rutherford had shown that α rays (or α *particles*) are just helium atoms carrying a charge of $+2e$. Assisted by Geiger and Marsden, Rutherford then began a systematic study of the scattering of α particles in matter. A schematic of Rutherford's scattering apparatus is shown in Fig. 11.2. Alpha particles from a radioactive source were confined to a narrow cone by a lead collimator. After scattering by a gold foil, the α particles were detected by observing the tiny flashes of light they produced upon striking a zinc sulfide screen. The detector could be rotated in order to measure the relative numbers of α particles scattered at various angles θ.

When observations of the scattered α particles were made, it was found,

Fig. 11.1 *Lord Rutherford, the key figure in unraveling the mysteries of radioactivity and in establishing the nuclear model of the atom. For his work on radioactivity, Rutherford was awarded the 1908 Nobel Prize (in chemistry).*

most unexpectedly, that about 1 α particle in 20,000 of those incident was turned completely around by a sheet of gold only 4×10^{-5} cm thick and emerged from the side facing the source. Rutherford commented: "It was quite the most incredible event that has ever happened to me in my life. It was almost as incredible as if you had fired a 15-inch shell at a piece of tissue paper and it came back and hit you."

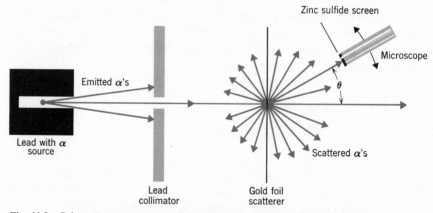

Fig. 11.2 *Schematic representation of the apparatus used by Rutherford to study the scattering of α particles. The entire apparatus was contained within an evacuated chamber in order to prevent the absorption of the α particles in air. Working under Rutherford's direction, Geiger and Marsden performed the experiments.*

It took several years (until 1911) for Rutherford to convince himself that he understood completely the meaning of the unexpected α-particle scattering that was observed at large angles of deflection. He concluded that the only way in which it is possible to account for the experimental results is to assume that the positive charge of the atom is concentrated in a small

ATOMS AND
MOLECULES

volume at the center of the atom instead of distributed throughout the atom as in a model that Thomson had proposed. Thus, Rutherford conceived the *nuclear* model of the atom.

In a collision between an α particle and an atom, the α particle should suffer very little deflection from the atomic electrons, but according to Rutherford, when the trajectory brings the α particle close to the massive nucleus, the intense electrical repulsion can cause a considerable change in the direction of motion of the α particle. Typical encounters between α particles and atoms are shown in Fig. 11.3.

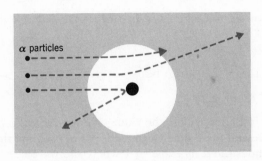

Fig. 11.3 *According to Rutherford's nuclear model of the atom, an α particle should usually pass through the atom with only little deflection, but occasionally the direction of motion of the α particle will bring it sufficiently close to the nucleus to cause a large-angle scattering.*

On the basis of his nuclear model, Rutherford proceeded to calculate in detail the way in which the intensity of scattered α particles should vary with the angle θ. These predictions were tested by Geiger and Marsden and were verified in every respect. It was conclusively demonstrated that atoms consist of nuclear cores of extremely small dimensions ($\sim 10^{-14}$ m) surrounded by atomic electrons.

11.2   *The Hydrogen Atom*

SPECTROSCOPIC RESULTS

In the middle of the 18th century it was discovered that when the light from flames is observed through a prism spectrometer there are certain discrete parts of the light spectrum that are more intense than the background continuum and stand out clearly as *lines*. Line spectra from a variety of sources (including the Sun) were investigated extensively during the latter half of the 19th century and *atomic spectroscopy* became a highly developed field of study. But there was no real understanding of the jumble of spectral lines that were observed.

In 1885 Johann Balmer (1825–1898), a Swiss music teacher with an interest in numbers, discovered that the wavelengths of the lines in the optical spectrum of hydrogen (Fig. 11.4) could be represented by a simple mathematical expression:

$$\frac{1}{\lambda} = R\left(\frac{1}{2^2} - \frac{1}{n^2}\right), \qquad n = 3, 4, 5, \ldots \tag{11.1}$$

where R is called the *Rydberg constant*, after the Swedish spectroscopist who made extensive investigations of atomic spectra. The value of R for the

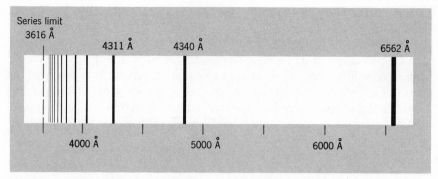

Fig. 11.4 *The Balmer series in the spectrum of hydrogen. The series limit (3646 A) corresponds to substituting* n = ∞ *in the Balmer formula (Eq. 11.1).*

hydrogen spectrum is now known to be

$$R = 10{,}967{,}758 \text{ m}^{-1} \text{ (hydrogen)} \tag{11.2}$$

although, of course, in Balmer's time the value was not known so precisely. How supremely successful the Balmer formula is in representing the hydrogen spectrum is demonstrated in Table 11.1 where the wavelengths calculated by Balmer are compared with those obtained from measurements made by the Swedish physicist, Anders Jonas Ångstrom (1814–1874). These four lines lie in the visible part of the spectrum. On the basis of his formula, Balmer also predicted that other lines should occur in the ultraviolet region. Indeed, in the spectra of some stars, as many as 50 lines in the Balmer series have been observed.

Table **11.1** *Comparison of Hydrogen Spectral Lines with Calculations from the Balmer Formula*

Line Designation	n	λ (Computed by Balmer)	λ (Observed by Ångstrom)
H_α	3	6562.08 Å	6562.10 Å
H_β	4	4860.80 Å	4860.74 Å
H_γ	5	4340.0 Å	4340.1 Å
H_δ	6	4101.3 Å	4101.2 Å

Although Balmer's formula was exceptionally accurate in reproducing the observations, no one understood why this should be so. It was almost 30 years before Niels Bohr provided the first glimmer of understanding.

BOHR'S INTERPRETATION OF ENERGY STATES

In 1913 Niels Bohr (1885–1962) took a bold and surprising step in an attempt to interpret the spectroscopic results for the hydrogen atom. He had accepted Rutherford's model of the atom with its nuclear core and outer electrons. According to classical theory, a system consisting of a massive, positively-charged core and light, negatively-charged electrons can be stable only if the electrons are in motion. Thus, an atom should be similar to a miniature solar system with a nuclear "Sun" and "planetary" electrons. The analogy would be expected to be quite good (after all, the electrical and gravitational forces both depend on $1/r^2$) were it not for the fact that classical theory also predicts that accelerating electric charges radiate energy in the form of electromagnetic waves. Therefore, the orbiting "planetary" electrons would be expected to lose their motional and electrical energy by radiation and rapidly fall toward the nucleus. A calculation based on classical electromagnetic theory shows that the electron in a hydrogen atom will radiate all of its energy in a small fraction of a second. But, of course, the atom does not do this—what is wrong with the classical model?

It was Bohr's audacious proposal that classical electromagnetic theory simply does not apply to an electron circulating in an orbit around a nucleus. Instead of attempting to *explain* the problem, Bohr simply *abolished* it! At the same time he reasoned that the series of discrete spectral lines implies that the photons are emitted only with certain definite frequencies (and, hence, certain definite energies). If the hydrogen atom can exist only in particular configurations, each with its own particular discrete energy, than as the atom changes from a configuration (or *state*) of higher energy $\mathcal{E}_n$ to one of lower energy $\mathcal{E}_{n'}$, a photon will be emitted with energy equal to the energy difference between the states:

$$\boxed{h\nu = \mathcal{E}_n - \mathcal{E}_{n'}} \tag{11.3}$$

In each of these energy states the electron moves in an orbit (in Bohr's view, a *classical* orbit) with a particular radius and velocity appropriate for that energy.

Bohr's new hypothesis was that an electron can exist in one of the discrete energy states in an atom, executing a radiationless (but otherwise classical) orbit; radiation occurs only when the electron makes a transition between two allowed orbits. These transitions occur spontaneously and continue until the atom is in its state of lowest allowed energy.

Having challenged the applicability of classical electromagnetic theory in the atomic domain, how was Bohr to sustain his claim?

BOHR'S ANGULAR MOMENTUM HYPOTHESIS

In order to justify his interpretation of the hydrogen spectrum in terms of radiations accompanying the transitions between allowed energy states, Bohr sought to calculate the energies of these states. He was able to obtain a set of discrete allowed states only by making the drastic assumption that *angular momentum is quantized.* By specifying that the angular momentum must be an integer multiple of $h/2\pi$, Bohr was finally able to derive the Balmer formula. (The combination $h/2\pi$ is used so frequently in atomic theory that it is given a special symbol: $h/2\pi = \hbar$.) Bohr's condition is, therefore (see Eq. 3.14),

$$L = m_e v r = n\hbar, \qquad n = 1, 2, 3, \ldots \tag{11.4}$$

n is called the *principal quantum number* for the particular state.

Example **11.1**

Show that Bohr's proposal of quantized angular momentum leads to Balmer's expression for the wavelengths of the lines in the hydrogen spectrum.

The electrical force between the orbiting electron and the nucleus in the hydrogen atom is

$$F_E = -k\frac{qQ}{r^2} = -k\frac{(-e)(+e)}{r^2} = k\frac{e^2}{r^2}$$

Since the nucleus is much more massive than the electron, we can consider the nucleus to remain in an essentially fixed position as the electron executes its orbit. The force on the electron is equal to its mass multiplied by its acceleration, and the acceleration is just the *centripetal* acceleration, $a_c = v^2/r$; thus,

$$F_E = m_e a_c = \frac{m_e v^2}{r}$$

Equating the right-hand sides of these two expressions for F_E and solving for v, we find

$$v = \sqrt{\frac{ke^2}{m_e r}}$$

Bohr's hypothesis of quantized angular momentum (Eq. 11.4), when combined with the expression for v, leads to

$$L = m_e v r = m_e r \times \sqrt{\frac{ke^2}{m_e r}} = e\sqrt{km_e r} = n\hbar$$

Squaring the last part of this equation and solving for r, we have

$$r_n = \frac{n^2 \hbar^2}{km_e e^2}$$

where we have attached a subscript n to the radius in order to indicate that this is the value for a particular value of n ($= 1, 2, 3, \ldots$).

Next, for the total energy of the nth orbit, we write, using the equation above for v,

$$\mathcal{E}_n = KE + PE = \frac{1}{2}m_e v^2 - k\frac{e^2}{r_n} = \frac{1}{2}m_e\left(\frac{ke^2}{m_e r_n}\right) - k\frac{e^2}{r_n} = -\frac{1}{2}k\left(\frac{e^2}{r_n}\right)$$

Substituting for r_n, we have, finally,

$$\mathcal{E}_n = -\frac{1}{2}\left(\frac{k^2 m_e e^4}{\hbar^2}\right)\left(\frac{1}{n^2}\right), \qquad n = 1, 2, 3, \ldots$$

Thus, by using the quantization condition on the angular momentum, Bohr succeeded in obtaining a set of discrete energy states. The value of the energy for a particular state depends on the value of the principal quantum number n for that state. For example,

$n = 1: \mathcal{E}_1 = -13.6$ eV
$n = 2: \mathcal{E}_2 = -3.39$ eV
$n = 3: \mathcal{E}_3 = -1.51$ eV
$n = 4: \mathcal{E}_4 = -0.85$ eV

The difference in energy between a state with principal quantum number n and one with n' is

$$\Delta \mathcal{E}_{nn'} = \mathcal{E}_n - \mathcal{E}_{n'}$$

$$= \frac{1}{2}\left(\frac{k^2 m_e e^4}{\hbar^2}\right)\left(\frac{1}{n'^2} - \frac{1}{n^2}\right)$$

If an electron makes a transition between these states (n to n', where $n > n'$), the wavelength of the radiation will be given by

$$\frac{1}{\lambda_{nn'}} = \frac{\nu_{nn'}}{c} = \frac{h\nu_{nn'}}{hc} = \frac{\Delta \mathcal{E}_{nn'}}{hc}$$

$$= \frac{1}{2}\left(\frac{k^2 m_e e^4}{\hbar^2}\right) \times \frac{1}{hc}\left(\frac{1}{n'^2} - \frac{1}{n^2}\right) = \frac{k^2 m_e e^4}{4\pi \hbar^3 c}\left(\frac{1}{n'^2} - \frac{1}{n^2}\right)$$

Substituting numerical values for the fundamental constants in this expression, Bohr found that the value of the quantity multiplying the terms in parentheses is very close to the experimental value of the Rydberg constant determined from the hydrogen spectrum (Eq. 11.2). By applying the angular momentum quantization condition to a classical model of the atom, Bohr was able to duplicate the results of measurements of the hydrogen atom spectrum.

Bohr's success in explaining the hydrogen spectrum on the basis of his half-classical, half-quantum model was not exactly hailed as a triumph. In fact, he was severely criticized for tampering with centuries of classical theories; even Bohr was at a loss to explain the fundamental significance of his curious mixture of classical dynamics and quantum hypotheses. It was more than 10 years before the development of the new quantum mechanics provided the proper explanation of Bohr's remarkable results.

HYDROGEN ENERGY STATES

By substituting various values of n and n' into his expression for the hydrogen atom wavelengths, not only was Bohr able to reproduce the results for the Balmer series of lines but he also predicted the existence of many other lines in the ultraviolet and infrared regions of the spectrum. Figure 11.5 shows the electron orbits in the Bohr model of the hydrogen atom together with some of the transitions in three of the series of lines that appear in the spectrum. Notice that each series consists of transitions that terminate at a particular orbit. That is, n' is the same for each line in a series. For the Balmer series, $n' = 2$ (see Eq. 11.1).

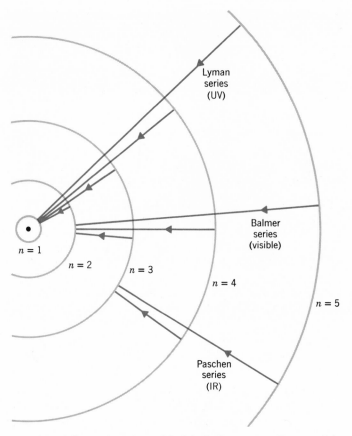

Fig. 11.5 *Orbits in the Bohr model of the hydrogen atom. Portions of three of the spectral series are shown.*

Example **11.2**

Calculate the binding energy of the hydrogen atom in its ground state.

Using the equation for $\mathcal{E}_n$ in Example 11.1 and substituting $n = 1$, we have

$$\mathcal{E}_1 = -\frac{k^2 m_e e^4}{2\hbar^2} = -\frac{2\pi^2 k^2 m_e e^4}{h^2}$$

$$= -\frac{2\pi^2 \times (9 \times 10^9 \text{ N-m}^2/\text{C}^2)^2 \times (9.1 \times 10^{-31} \text{ kg}) \times (1.60 \times 10^{-19} \text{ C})^4}{(6.6 \times 10^{-34} \text{ J-sec})^2}$$

$$= -2.18 \times 10^{-18} \text{ J} \times \left(\frac{1 \text{ eV}}{1.60 \times 10^{-19} \text{ J}}\right) = -13.6 \text{ eV}$$

This is the *total* energy of the state (and is *negative* because of our convention in which all energies of bound systems are negative). The *binding energy*, the energy that must be supplied to raise the total energy to *zero* and thereby release the electron, is 13.6 eV.

ATOMS AND
MOLECULES

240

Bohr's explanation of the hydrogen spectrum in terms of electrons orbiting with quantized angular momenta was tremendously successful in accounting for the observed spectral lines. But the quantization rule remained as an *ad hoc* hypothesis, not based on any deeper theory, until de Broglie made his famous proposal of the wave character of matter. Instead of a tiny, localized electron orbiting the nucleus, de Broglie's idea was that there is an electron *wave* encircling the nucleus. If the electron wave requires a distance equal to the circumference of the orbit (as specified by Bohr) to exactly complete an integer number of cycles, this means that the wave will join smoothly onto itself and constructive interference or self-reinforcement will result (see Fig. 11.6a). On the other hand, if we try to fit into a certain orbit a wave with a wavelength that is not an integer fraction of the circumference, *destructive* interference will result and the wave will rapidly damp to zero amplitude (see Fig. 11.6b). Therefore, each Bohr orbit contains an integer number of de Broglie electron wavelengths; the first Bohr orbit has a circumference equal to one electron wavelength, the second has a circumference equal to two wavelengths, and so forth, each orbit differing in circumference from the next by one wavelength. Each orbit has its own particular value of the wavelength because the velocity of the electron in each orbit is different.

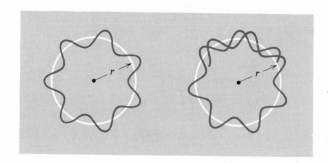

Fig. 11.6 (*a*) *Constructive interference results when an integer number of wavelengths are just fitted onto the circumference of an orbit so that this state of the system is maintained by self-reinforcement. (b) Destructive interference results when the integer wavelength condition is not satisfied and the state is rapidly damped to zero amplitude.*

De Broglie's proposal of the wave character of matter therefore explains in a simple and straightforward way the puzzling angular momentum quantization rule of Bohr. The merging of Bohr's and de Broglie's concepts provided much of the impetus that led within a short period of time to the development of the modern theory of atomic structure in quantum mechanical terms.

11.3 *Angular Momentum and Spin*

MAGNETIC SUBSTATES

Although Bohr's model of the hydrogen atom was capable of accounting for most of the lines in the hydrogen spectrum, Bohr was unable to extend the same ideas and apply them to the spectra of more complicated atoms.

The theory was clearly in need of modification. The next step was taken by the German theorist, Arnold Sommerfeld (1869–1951).

According to the Bohr model, the principal quantum number n specifies the angular momentum for the state by the relation $L = n\hbar$ (Eq. 11.4). That is, each orbit is described completely by the quantum number n and has a unique value of angular momentum, $n\hbar$. In 1915 Sommerfeld showed that there were additional values of the angular momentum available for each n. In Sommerfeld's extension of the Bohr model, the principal quantum number n is retained as a measure of the *energy* of the state, and a new quantum number l is introduced to specify the *angular momentum*:

$$L = l\hbar, \qquad l = 0, 1, 2, \ldots, n - 1 \tag{11.5}$$

where l can have only positive values from zero to $n - 1$. For convenience in discussions, each angular momentum state is given a letter designation:

$l = 0$	S state
$l = 1$	P state
$l = 2$	D state
$l = 3$	F state
$l = 4$	G state

For example, states with $l = 0, 1$, and 2 (that is S, P, and D states) are allowed for the principal quantum number $n = 3$. These states are labeled $3S$, $3P$, and $3D$, where the number preceding the letter indicates the value of n. The various allowed states for $n = 1$ to 5 are listed in Table 11.2.

Table **11.2** *Allowed States and Their Designations*

n	$l = 0$ (S)	$l = 1$ (P)	$l = 2$ (D)	$l = 3$ (F)	$l = 4$ (G)
1	1S				
2	2S	2P			
3	3S	3P	3D		
4	4S	4P	4D	4F	
5	5S	5P	5D	5F	5G

An orbiting atomic electron acts as a tiny current loop and produces a magnetic field (see Fig. 7.9). If the atomic electron and its magnetic field are placed in an external magnetic field (Fig. 11.7), there will be an interaction between the two fields. By observing the spectral lines emitted by atoms when in a magnetic field, an important conclusion has been reached. The N-S axis of the electron orbit (which is the same as the direction of the angular momentum vector **L**) cannot point in any arbitrary direction with respect to the field direction; instead, the direction of **L** is restricted to certain discrete directions relative to **B**. The allowed directions of **L** are those for which the component of **L** in the direction of **B** (the z-direction) is an integer multiple of $\hbar$:

$$L_z = m_l\hbar, \qquad m_l = l, l - 1, \ldots, 0, 1, \ldots, -l + 1, -l \tag{11.6}$$

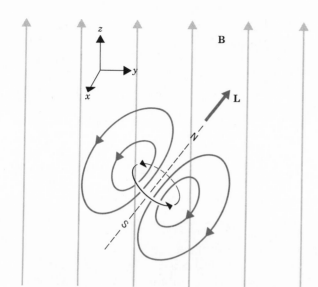

Fig. 11.7 *The magnetic field produced by an orbiting atomic electron interacts with an external field **B**. Only certain discrete directions of **L** relative to **B** are allowed.*

Angular momentum states are therefore further divided into *magnetic substates* specified by the *magnetic quantum number* m_l.

Since the component of **L** can be either *along* the direction of **B** or *opposite* to it, m_l can have both positive and negative values; $m_l = 0$ is always possible (**L** perpendicular to **B**) and the maximum and minimum values of m_l are $+l$ and $-l$, respectively. Thus, there are always $2l + 1$ allowed values of m_l (see Table 11.3).

Table **11.3** *Magnetic Substances*

State	l	m_l	$2l + 1$
S	0	0	1
P	1	+1, 0, −1	3
D	2	+2, +1, 0, −1, −2	5
F	3	+3, +2, +1, 0, −1, −2, −3	7

SPIN

In spite of the obvious great success of the Bohr-Sommerfeld picture of atomic structure, there were still many unexplained facts even for relatively simple systems. For example, certain spectral lines that were expected to be single were found, on close examination, actually to be *doublets* spaced apart by only a very small energy.

In 1925, Goudsmit and Uhlenbeck argued that this effect indicates that an electron possesses angular momentum quite apart from the angular momentum that arises from orbital motion. In classical terms, we can picture an electron as a spinning, charged ball—the mechanical spin produces an angular momentum. This classical model has no meaning within the framework of quantum theory (where we speak only of the *intrinsic* angular momentum in exactly the same way that we speak of the intrinsic *mass* or the intrinsic *charge* of an electron). Nevertheless, the classical model is a convenient picture and is often used; indeed, it is customary to refer to intrinsic angular momentum simply as *spin*. The spin of an electron is $S = \frac{1}{2}\hbar$.

Measured values of angular momenta can differ only by integer multiples of the fundamental unit of angular momentum, $\hbar$. This is the reason that both L and L_z are proportional to $\hbar$. Similarly, S_z, the projection of the spin angular momentum $\mathbf{S}$ on a preferred axis, can take on only values that differ by integer multiples of $\hbar$. But the magnitude of S_z cannot exceed $S = \frac{1}{2}\hbar$. Therefore, there are only *two* allowed values for S_z: $+\frac{1}{2}\hbar$ and $-\frac{1}{2}\hbar$. That is,

$$S_z = m_s\hbar, \qquad m_s = +\frac{1}{2}, -\frac{1}{2} \qquad (11.7)$$

Notice that $S_z = 0$ is not allowed.

By this stage in the development of the theory of atomic structure, the original Bohr theory had been modified almost beyond recognition. No longer were atomic electrons thought to be orbiting "planets" described by a single quantum number n. Instead, it was recognized (but not completely understood) that the wave property of electrons is important in atoms; furthermore it was known that *four* quantum numbers are required to specify the state of an atomic electron (see Table 11.4).

Table **11.4**   *Quantum Numbers of Atomic States*

Property	Quantum Number	Allowed Values
Energy	$n \left(E \propto -\dfrac{1}{n^2} \right)^{\dagger}$	1, 2, 3, . . .
Angular momentum	$l \quad (L = l\hbar)$	0, 1, 2, . . . , n-1
Projection of angular momentum	$m_l \quad (L_z = m_l\hbar)$	$-l$ to $+l$ ($2l + 1$ values)
Spin projection	$m_s \quad (S_z = m_s\hbar)$	$+\frac{1}{2}, -\frac{1}{2}$

†See Example 11.1

THE DEVELOPMENT OF QUANTUM THEORY

By the mid-1920s it was generally recognized that the Bohr-Sommerfeld ideas of atomic structure, including as they did both classical and quantum concepts, left much to be desired in terms of a complete and satisfying physical explanation of the properties of atoms. In 1925 and 1926 there emerged a new view of atomic processes that was based, not on a description of electron orbits and "jumping" electrons, but on the *wave* properties of electrons. The classical idea of orbits was abandoned and in its place came the *wave mechanics* or *quantum theory* of elementary processes that was developed during the 1920's by Heisenberg, Schrödinger, Born, Pauli, Dirac, and others.

The birth of quantum theory was a gigantic step forward in our understanding of Nature; it is even more remarkable when it is realized that all of the crucial developments took place in such a short period of time. Quantum theory is certainly an equal, if not a greater tribute to the powers

ATOMS AND
MOLECULES

of the human intellect than was Newton's formulation of the law of the universal gravitation and his explanation of planetary motion.

If we surrender the classical idea of electron orbits, how are we to understand the various energy states of the hydrogen atom? Schrödinger approached this problem by writing an equation that was quite similar to the equation that describes the propagation of mechanical waves and in which he included a term to represent the effect of the electrostatic potential energy of the electron in the field of the nuclear proton. The solution of this equation showed that there are certain discrete energies allowed for the system. These energies, which emerge in a natural way from the Schrödinger equation, correspond precisely to the energies in the Bohr theory but there are no "orbits" for the electron. Instead, each energy state has associated with it a wave function which represents the amplitude of the electron wave at any point in the vicinity of the nucleus. The square of this amplitude is proportional to the probability for finding the electron at any particular position.

11.4 *The Exclusion Principle and Atomic Shell Structure*

SYSTEMATICS OF THE PROPERTIES OF THE ELEMENTS

The Schrödinger-Heisenberg quantum mechanics of 1925 proved to be enormously successful in explaining the spectra of hydrogen and other one-electron systems. The introduction of spin permitted an understanding of some of the results for more complicated atoms. But at this stage it could not be claimed that the structure of atoms containing many electrons was understood in detail.

It was known, from the systematic study of atomic radiations, that there is a regular progression in the number of atomic electrons as one passes from element to element and that this electronic charge is balanced by an equal positive charge on the nucleus. Thus, hydrogen has one orbital electron and a charge of $+e$ on the nucleus; helium has two orbital electrons and a charge of $+2e$ on the nucleus. The number of electrons in the neutral atom (or, equivalently, the number of nuclear charges) is called the *atomic number* of the element and is denoted by Z.

It was also known that many of the physical and chemical properties of the elements could be organized in a systematic way and presented in the form of the *periodic table of the elements* (Fig. 11.8). In this table the elements are arranged according to *groups* and *periods*, with the members of each group having similar properties. Thus, the elements Li, Na, K, Rb, Cs, and Fr are all similar to hydrogen in that they participate in chemical reactions as if they have only a single effective electron (called a *valence* electron). For example, the Group I elements combine readily in a one-to-one fashion with the Group VII elements to form such compounds as HF, HBr, NaCl, NaF, KCl, KBr. The Group O elements, the so-called *noble* or *inert gases,* do not readily combine with other elements; these gases have no valence electrons.

The periodic table represents the cyclic behavior of many chemical and physical properties of the elements. Each of these cycles ends with a noble gas. Thus, the various periods terminate at the atomic numbers $Z = 2, 10, 18, 36, 54,$ and 86. This periodicity is revealed in a striking way by the

11.4
THE
EXCLUSION
PRINCIPLE
AND ATOMIC
SHELL
STRUCTURE

245

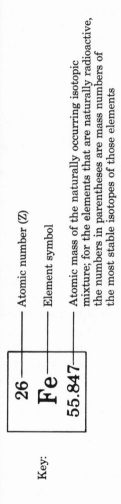

PERIODIC TABLE OF THE ELEMENTS

Fig. 11.8 *Periodic table of the elements.*

ionization energies of the elements. (This *ionization energy* is the minimum energy required to remove an electron from an atom and convert it into a singly-charged ion.) Figure 11.9 shows that the ionization energy tends to be quite large for the noble gases and to be quite low for the element with the next higher atomic number (a Group I element with an easily removed valence electron). There is a more-or-less uniform increase in the ionization energy as we proceed across any given period from Group I to Group O.

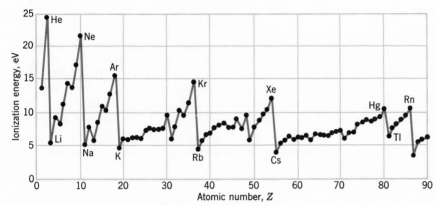

Fig. 11.9 *Ionization energies of the elements as a function of the atomic number.*

In spite of the obvious importance of the systematic behavior of elements with regard to both physical and chemical properties, there was no clue in either the old Bohr-Sommerfeld model or in the early quantum theory as to the reason. All that could be said was that electrons seemed to exist in layers or *shells,* with each successive shell ending or closing with a noble gas so that there are no electrons (*valence* electrons) available to participate in chemical reactions. The elements at the beginning of each shell (the Group I elements) have one valence electron; the Group II elements have two valence electrons; and so on. The significance of this electronic shell structure and the meaning of the shell-closure numbers 2, 10, 18, 36, 54, and 86 remained a mystery until the solution was given in a simple and elegant way by Wolfgang Pauli.

THE EXCLUSION PRINCIPLE

The key to the problem of atomic shell structure was discovered by Pauli in 1925. The closing of atomic shells implies that an arbitrarily large number of electrons cannot be placed in a given shell. Pauli realized that such a restrictive effect must have a truly fundamental cause, and his solution to the problem was the formulation of the following principle, known as the *exclusion principle:*

No two electrons in an atom can have identical sets of quantum numbers.

That is, if one atomic electron is in a certain quantum state defined by a set of quantum numbers, n, l, m_l, and m_s, then other electrons in that atom

11.4
THE
EXCLUSION
PRINCIPLE
AND ATOMIC
SHELL
STRUCTURE

247

are excluded from that particular quantum state. It is most remarkable that the details of atomic structure can follow from a principle so simply stated. But such is the beauty of Nature's way.

How many states are available to an electron in an atom? We shall limit our considerations now to the *ground states* of neutral atoms; that is, the Z electrons in the atom are arranged in the way that produces the *minimum* total energy for the system. For $n = 1$, only $l = 0$ is possible and, therefore, only $m_l = 0$ is possible. But there are two possible spin states, $m_s = +\frac{1}{2}$ and $m_s = -\frac{1}{2}$. Therefore, two electrons exhaust the $n = 1$ states and the first shell is filled for $Z = 2$ (helium), as shown in Fig. 11.10. The first shell, which contains only the two $n = 1$, S electrons is called the *K shell*.

In order to form lithium ($Z = 3$), we must add the third electron in a state with $n = 2$ and with lithium we begin the L shell. For $n = 2$ we have available two 2S states ($l = 0, m_l = 0, m_s = \pm\frac{1}{2}$) and six 2P states (two states from $l = 1, m_l = +1, m_s = \pm\frac{1}{2}$, two states from $l = 1, m_l = 0, m_s = \pm\frac{1}{2}$, and two states from $l = 1, m_l = -1, m_s = \pm\frac{1}{2}$). Therefore, the L shell has a total of 8 possible electron states and this shell consists of the 8 elements from $Z = 3$ (lithium) through $Z = 10$ (the noble gas, neon), as indicated in Fig. 11.10.

HIGHER SHELLS

For the third and higher shells the situation is complicated by the fact that the inner electrons effectively shield a portion of the nuclear charge so that the outer electrons do not experience the full force of the $+Ze$ charge on the nucleus. Furthermore, the importance of this effect is dependent on the angular momentum of the outer electrons (because the electrons with low angular momentum "dip" into the inner electron shells whereas those with high angular momentum do not). The shielding and angular momentum effects give rise to the occurrence of the *transition elements* in the middle of the periodic table and to the *lanthanide* and *actinide* series of rare-earth elements (see Fig. 11.8).

11.5 X Rays

INNER-SHELL TRANSITIONS

Most of the properties of atoms—chemical, electrical, magnetic, optical, etc.—depend upon the configurations of the outermost electrons. Only in the event of a very energetic disturbance are the tightly bound inner electrons involved in the process. The reason is easy to see on the basis of energetics; it requires, for example, only 7.4 eV to remove the outermost electron from a lead atom ($Z = 82$), but an energy of 88 keV or 88,000 eV is necessary to remove one of the K electrons.

If sufficient energy is supplied to an atom by collision with a fast electron (as in an X-ray tube) or by irradiation with an energetic photon, then it is indeed possible to remove one of the inner K electrons (see Fig. 11.11a). The atom will not remain long in this condition with a vacancy in its K

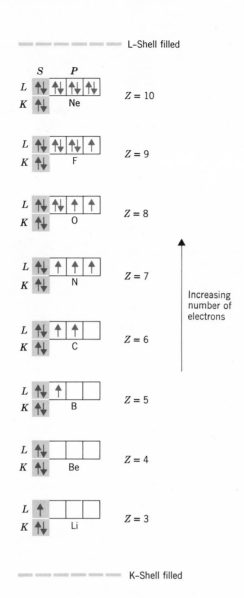

$Z = 10$ Ne

$Z = 9$ F

$Z = 8$ O

$Z = 7$ N

Increasing number of electrons

$Z = 6$ C

$Z = 5$ B

$Z = 4$ Be

$Z = 3$ Li

K-Shell filled

$Z = 2$ He

$Z = 1$ H

Fig. 11.10 *Filling of the first two shells of atomic electrons. In the first column are the S states (1S and 2S) and in the next three columns are the P states (1 = 1, m_1 = +1, 0, −1). The boxes represent the magnetic substates; each substate contains two spin states, $m_s = \pm\frac{1}{2}$, which are represented by the arrows. The K shell is filled at helium and the next 8 electrons must be placed in the L shell which is completed at $Z = 10$, neon.*

shell. It is energetically more favorable for an electron in a higher shell to make a transition and occupy the K-shell vacancy (Fig. 11.11*b*). It is most likely that an L electron will make this transition, emitting an energetic photon (an X ray) in the process. But then there is a vacancy in the L shell which is filled by an electron from one of the higher shells. Eventually, after

11.5
X RAYS

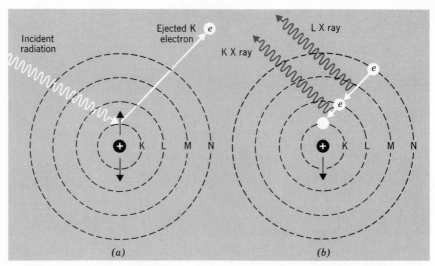

Fig. 11.11 *Schematic representation of the production of X rays. (a) A high-energy photon is incident on an atom; it penetrates to the innermost shell and ejects one of the two K electrons. (b) One of the L electrons then makes a transition, filling the vacancy in the K shell and emitting a K X ray in the process. Subsequently, an N electron makes a transition, filling the new vacancy in the L shell and emitting an L X ray. Finally, the vacancy in the outermost shell is filled by the capture of a free electron from the surroundings and the atom is again electrically neutral.*

this cascading of electrons and the emission of a series of X-ray photons, a free electron from the surroundings will be captured into the outer shell and return the atom to an electrically neutral condition.

X rays are easily produced by allowing electrons to move through a potential difference of 10,000–20,000 volts or so within an evacuated tube and then strike a metal electrode. The X rays produced in such devices are much used in medical diagnostic and therapeutic work. Because X rays have high energies they can penetrate into the body and damage the living tissue. Therefore, caution should be exercised in dealing with any equipment (including high-voltage television sets) that is capable of producing X rays; see Section 12.4.

11.6   *Molecules*

IONIC BINDING

We are all familiar with the chemical compound *sodium chloride* (NaCl)—it is just ordinary *salt,* but it is typical of a large class of simple molecules. The molecule of sodium chloride consists of two atoms, one of the metal sodium and one of the gas chlorine. How are these two dissimilar atoms—a metal and a gas—bound together to form a stable substance such as NaCl? If we refer to the periodic table (Fig. 11.8), we see that sodium ($Z = 11$) is the first element in Period 3 and thus has one electron outside the closed L shell (that is, Na is a Group I element). Chlorine ($Z = 17$) is the Period 3, Group VII element and therefore has 7 electrons outside the closed L shell or, equivalently, lacks one electron to fill the third shell.

The single valence electron of sodium is relatively easy to remove; it requires only 5.1 eV of energy to detach this electron and form a positively-charged sodium ion, Na$^+$ (see Fig. 11.12a). An atom of chlorine, on the other hand, has an affinity for electrons and, if provided with a free electron, will absorb this electron into its outer shell, thus completely filling this shell. This process forms a negatively-charged chlorine ion, Cl$^-$, and *releases* 3.7 eV of energy (see Fig. 11.12b).

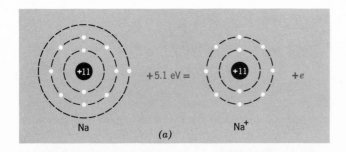

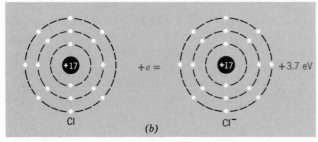

Fig. 11.12 (a) *It requires 5.1 eV of energy to detach the single valence electron of sodium and form the ion, Na$^+$. (b) A chlorine atom will acquire an electron from its environment to fill the third shell and form the ion, Cl$^-$; in the process, 3.7 eV of energy is released. The two ions, Na$^+$ and Cl$^-$, attract one another and form the ionic salt Na$^+$Cl$^-$, (or, simply, NaCl).*

The removal of an electron from sodium and the acquisition of an electron by chlorine are complementary situations. Sodium and chlorine can therefore exist together as a chemical compound, NaCl, in which the sodium electron is used to complete the outer electron shell of chlorine. But how can such a compound be stable since it requires 5.1 eV to remove the sodium electron and only 3.7 eV is gained by forming Cl$^-$? The answer lies in the fact that the electron transfer produces the ions Na$^+$ and Cl$^-$ and these ions are than attracted toward one another by the electrostatic force. When the centers of the two ions are separated by a distance of about 11 Å, the electrostatic potential energy of the system has contributed the requisite 1.4 eV to effect the electron transfer. In fact, the electrostatic attraction pulls the ions even closer together and binds them more tightly. But the ions cannot approach more closely than a certain small distance (which turns out to be 2.4 Å) with all of the electrons in the lowest possible energy state; if they did, two electrons with the same set of quantum numbers would be occupying the same region of space and this is prohibited by the exclusion principle. The only alternative would be for one or more electrons to be raised into a higher energy state, and under such conditions the molecule would no longer be a bound system. At the equilibrium separation of 2.4 Å imposed by the exclusion principle, the binding energy of the ionic molecule is 5.5 eV.

The binding together of atoms by virtue of the electrostatic attraction between ions formed by the *transfer* of an electron from one atom to the other is called *ionic binding*. In addition to NaCl, many other molecular compounds are bound in this way, for example. NaBr, KCl, RbI, and LiF. The compound MgO is produced by the transfer of *two* electrons from magnesium to oxygen, forming Mg^{++} and O^{--}. The compound Na_2S is produced by the transfer of one electron from each of the two sodium atoms to the sulfur atom, forming $2\,Na^+$ and S^{--}.

COVALENT BINDING

Ionic binding results from the transfer of one or more electrons from one atom to another; *covalent binding* results from the *sharing* of one or more electrons by the atoms. The simplest molecule that exhibits covalent binding is the hydrogen molecule, H_2. Because there are *two* electrons in the hydrogen molecule, the exclusion principle must be obeyed and the two electrons in the molecule cannot have the same set of quantum numbers. In the lowest energy state (the ground state), the values of n, l, and m_l are the same for the two electrons. Therefore, the electrons must have different values of m_s; that is, the electron spin vectors must be in *opposite* directions. Figure 11.13*a* shows the "clouds" of electron probability for the H_2 molecule (spins opposite). It is apparent that each electron has a high probability of being found *between* the two protons. In this condition each electron partially shields one of the protons and attracts both of the protons. The result is a bound molecule. Figure 11.13*b*, on the other hand, shows the situation in the event that the spins are parallel. The exclusion principle prevents the two electrons, which now have all four quantum numbers the same, from occupying the same region of space. Consequently, the probability cloud of each electron is concentrated on the opposite side of the proton, there is no mutual attraction, and no binding results.

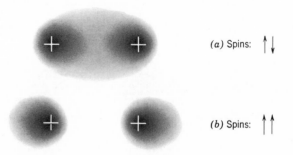

Fig. 11.13 *The effect of the exclusion principle on the hydrogen molecule. (a) Bound H_2 molecule with spins opposite; binding is due to the concentration of the electron probability density cloud between the protons. (b) H_2 "molecule" with spins parallel; the electron clouds lie primarily outside of the protons, and there is no molecular binding.*

(a) Spins: ↑↓

(b) Spins: ↑↑

Among the large number of other molecules that are formed by covalent binding are water (H_2O) and ammonia (NH_3). Figure 11.14 illustrates schematically how these molecules employ covalent bonds. The box diagrams for oxygen and nitrogen, showing the electron spin states in the K

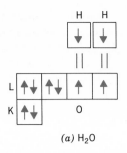

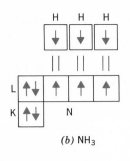

(a) H₂O

(b) NH₃

Fig. 11.14 *Schematic representations of the covalent bonding in (a) water and (b) ammonia. The double lines indicate the bonds formed by the sharing of two electrons.*

and L shells are the same as those in Fig. 11.10. Oxygen has two unpaired electrons in the L shell while nitrogen has three such electrons. Each of these electrons can effect a covalent bond with a hydrogen atom, forming H_2O and NH_3. The spatial structures of these molecules are illustrated in Fig. 11.15. Of course, the molecules are not rigid as suggested in the diagrams, which are only schematic, but the *average* orientations of the electron clouds can be measured and have the directions shown.

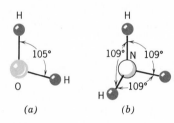

(a)

(b)

Fig. 11.15 *(a) Representation of the water (H_2O) molecule. The straight lines indicate covalent bonds.*
(b) Representation of the ammonia (NH_3) molecule.

CARBON BONDS

The electronic configuration of the carbon atom in its ground state, Fig. 11.16a, indicates that there are two unpaired P electrons in the L shell. We would expect, therefore, that carbon atoms would participate in the forma-

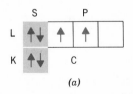

(a)

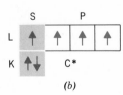

(b)

Fig. 11.16 *(a) The electronic configuration of the ground state of the carbon atom. Very little energy is required to break the 2S electron pair and promote one of the electrons into the 2P sub-shell where it remains unpaired. In this excited atomic state (b) there are four unpaired electrons, all of which participate in covalent bonding.*

tion of molecules by contributing two electrons toward covalent bonds. It is found, however, that carbon almost always appears in molecular structures with *four* equivalent covalent bonds. The reason that all four of the L electrons in carbon participate in molecular bonding, instead of just the two unpaired electrons of the ground-state configuration, is the following. By the addition of only a small amount of energy (about 2 eV) to the carbon atom ground state, it is possible to break the 2S electron pair and promote one of the electrons into the 2P subshell (Fig. 11.16b). Thus, there are *four* unpaired electrons available for bonding if 2 eV can be supplied to the atom.

11.6
MOLECULES

Actually, the energy is supplied in the bonding process itself because the energy gained by making four covalent bonds, instead of two, more than compensates for the energy expended in breaking the 2S pair. Three of these bonds are made by P electrons and one by an S electron. This type of equivalent four-electron bonding of the carbon atom to other atoms is called SP^3 bonding or *hybridization*.

HYDROCARBON MOLECULES

The simplest of the *organic* molecules (molecules that are found in living things) are those that consist entirely of carbon and hydrogen—the *hydrocarbons*. Many different combinations of C and H are possible; thousands are known and they occur in gaseous, liquid, and solid forms at room temperature. The common commercial fuels, gasoline and natural or liquid (LP) gas, are mixtures of hydrocarbons. The simplest of the hydrocarbons is methane, CH_4, in which each of the four available carbon bonds is utilized to bind a hydrogen atom in the molecule. The structure of methane is shown schematically in Fig. 11.17a where the short dashes each represent a covalent bond involving two electrons. In the ethane molecule (Fig. 11.17b), one of the carbon bonds is utilized to attach another carbon atom, resulting in C_2H_6. More complicated, long-chain hydrocarbon molecules are formed by adding additional CH_2 units.

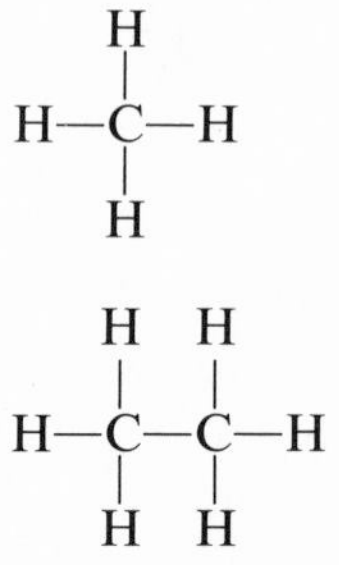

(a) H—C—H   Methane, CH_4

(b) H—C—C—H   Ethane, C_2H_6

Fig. 11.17 *The first two molecules in the series of hydrocarbon chain molecules.*

The fourth member of the group of long-chain hydrocarbons is *butane*, C_4H_{10}, shown in Fig. 11.18. Notice that the carbon atoms do not lie along a straight line—it is energetically more favorable for the atoms to form a kinked line.

In addition to *chain* molecules, carbon and hydrogen atoms can combine to form *ring* molecules, such as that of the compound *benzene*, C_6H_6, shown in Fig. 11.19. Notice that in this molecule the carbon atoms are connected by alternating *single* bonds (that is, *two* shared electrons) and *double* bonds (that is, *four* shared electrons).

11.7 *Solids*

ENERGY BANDS

ATOMS AND MOLECULES

Figure 11.20 shows the electrostatic potential in the vicinity of an isolated atom of lithium. The horizontal line labeled 0 indicates zero potential energy

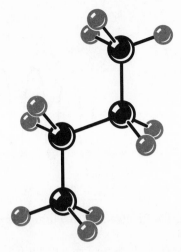

Fig. 11.18 *The molecular structure of* butane, C_4H_{10}.

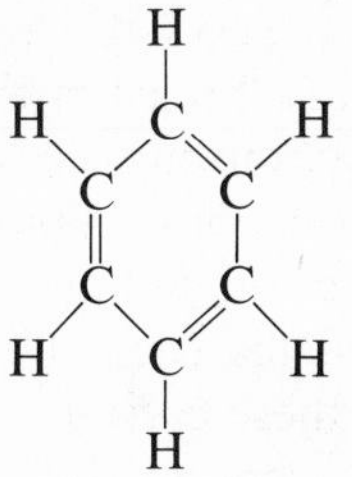

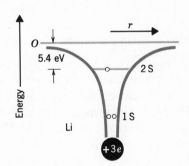

Fig. 11.19 *The ring compound,* benzene, C_6H_6.

and the two lower lines indicate the 1S and 2S energy states. Two electrons are located in the 1S state and one in the 2S state. The 2S electron is bound by 5.4 eV; that is, the ionization energy of an isolated lithium atom is 5.4 eV.

In solid lithium the atoms form a regular array or *crystal*. The net electrostatic field at any point in the crystal is the sum of all of the individual

Fig. 11.20 *Schematic potential energy diagram for an isolated lithium atom. The outermost electron (i.e., the 2S electron) is bound by 5.4 eV; it requires $\sim$75 eV to remove one of the 1S electrons and an additional $\sim$120 eV to remove the second 1S electron.*

atomic fields. Consequently, the potential between the atoms never rises to zero potential. In fact, the potential at a particular point between atoms in a crystal is reduced substantially below the potential at the same distance from an isolated atom (Fig. 11.21). The reduction is so pronounced in the crystal that the 2S electron, which was *bound* in the isolated atom, no longer encounters a potential barrier that is sufficient to constrain it to the vicinity

11.7
SOLIDS

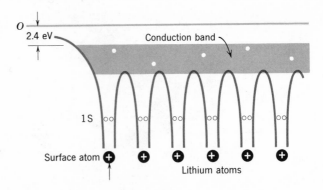

Fig. 11.21 *When lithium atoms are assembled in a crystal, the reduction of the potential between the atoms frees the 2S electrons which then partially fill the conduction band. (The energies are not to scale; a 1S electron is actually bound by 65 eV.) The electrons in the conduction band, shown schematically in the diagram, are not really localized; the electron wave functions extend throughout the crystal.*

of any particular atom. The 2S electrons in a lithium crystal are *free* electrons or *conduction* electrons.

The conduction electrons "belong," not to individual atoms, but to the crystal as a whole. That is, the 2S electron wave functions are not localized but extend throughout the crystal. The 2S energy state of the isolated lithium atom becomes an energy "state" of the crystal. If there are N atoms in the crystal, there are N electrons in this "state." But we know that the exclusion principle does not permit more than two electrons to exist in any single energy state. Therefore, the atomic 2S state must expand into a series of closely spaced crystal states, each of which can accommodate two electrons. The spacing between these crystal states is so small[1] that the distribution of states is essentially continuous. As a result, the discrete atomic state becomes, in the crystal, an energy *band* (see Fig. 11.21). This band is the *conduction band* and the electrons that exist in this band are the *conduction electrons*.

The interesting physical properties of an element or compound in bulk form (for example, electrical resistance, thermal conductivity, magnetic properties) are determined in large measure by the details of the energy band structure. We shall investigate next how the crystal energy bands determine the electrical conductivity of the material.

BAND THEORY OF CONDUCTORS AND INSULATORS

In general, all the energy states of an atom appear as bands in a crystal, including all of those higher states that are empty when the atom is in its ground state. Thus, in a lithium crystal there are (unfilled) bands corresponding to the 2P, 3S, 3P, 3D, etc. states. The discrete atomic states can appear in the crystal as identifiable bands (Fig. 11.22*b*) or as overlapping bands in which it is no longer possible to specify from which subshell the electrons originated (Fig. 11.22*c*). Overlapping always occurs for the highest energy bands but the lower states usually remain as individual bands in the crystal. The details of the crystal lattice structure determine the excitation energy at which the overlapping begins.

The filling of the energy bands by electrons follows the same prescription

[1] These states are spread over an energy of a few eV, and a total of $\frac{1}{2}N$ states are required to accommodate N electrons. For a 1-cm³ crystal, N is of the order of 10^{22}, so the spacing between the states is of the order of 10^{-22} eV.

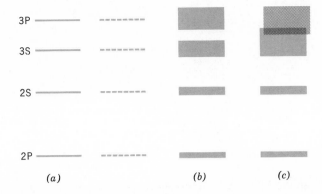

3P

3S

2S

2P

(a)   (b)   (c)

Fig. 11.22   *The discrete atomic energy states (a)
correspond, in a crystal, to either individual bonds
(b) or to overlapping bands (c). (The 1S state
always lies at a much lower energy than the other
states and so is not shown in this diagram.)*

as does the filling of atomic energy states. The 2S atomic state; for example,
can accommodate two electrons and in a crystal consisting of *N* atoms, the
2S band can contain 2N electrons; the 3P band (if distinct from the 3S and
3D bands) can contain 6N electrons; and so forth. If a band is completely
filled, no electron in this band can be given any additional energy unless
it is given a sufficient amount to raise it to an unoccupied state in a higher
band (Fig. 11.23*a, b*). Depending on the positions of the various bands, the
amount of energy required to raise an electron from one band to another
may be 5–10 eV. On the other hand, if the highest energy band that contains
any electrons is only partially filled (for example, the 2S band in lithium
which contains N electrons and thus is half filled), there is an extremely
large number of energy states *within* the band that are accessible to these
electrons. Thus, an electron in a partially filled band can be given essentially
any amount of additional energy as long as the total is less than the maximum energy allowed for the band (Fig. 11.23*c*).

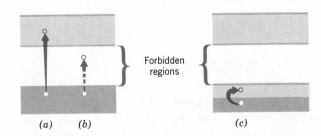

Forbidden
regions

(a)   (b)   (c)

Fig. 11.23   *(a) An allowed transition of an
electron from a filled band into an empty band;
this type of transition usually requires 5–10 eV.
(b) The transition of an electron from a filled
band into the forbidden region is not allowed.
(c) In a partially filled band an electron can make
a transition into any unoccupied state that lies
within the band; such transitions ordinarily
involve only very small amounts of energy.*

If the highest occupied band is only partially filled, the electrons in this
band can be made to drift in a particular direction by the application of
an external electric field. The increase in energy brought about by this
motion (since it is small) can be accommodated by the available energy
states within the band. Materials that have partially filled bands can therefore conduct electricity and are called *conductors* (Fig. 11.24*a*).

If the highest occupied band is completely filled, increases in the energies
of the electrons in that band are not allowed and the application of an
electric field will not result in the flow of electrons. Such materials resist
the flow of electricity and are called *insulators* (Fig. 11.24*b*).

In most insulators it requires an appreciable increase in energy (5–10 eV)

11.7
SOLIDS

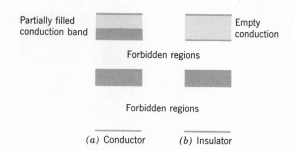

Partially filled
conduction band

Empty
conduction

Forbidden regions

Forbidden regions

(a) Conductor   (b) Insulator

Fig. 11.24 *A conductor (a) is characterized by a partially filled band and an insulator (b) by a completely filled band above which is a forbidden region and, still higher, an empty band.*

to raise an electron from the filled band across the forbidden energy region and into the empty conduction band. For example, in the diamond form of carbon there is a 5-eV energy gap between the filled 2P band and the lowest empty band (Fig. 11.25). What electric field strength is necessary to raise an electron from the 2P band into the empty band? In order to answer

2P Empty conduction band

5 eV

2P

2S

1S

C

Fig. 11.25 *The energy bands of carbon in the form of diamond (not to scale). Because of the relatively large energy gap between the filled 2P band and the empty band, diamond is a good insulator.*

this question we must first realize that no *real* crystal is as perfect as the ideal crystal structures we have been discussing. There are always small amounts of impurities present in real crystals and there are always small imperfections in the lattice structure. These departures from the ideal crystal form prevent the electrons from moving unimpeded through the crystal. In fact, in even the purest crystals that have been made, an electron can travel only about 10^{-8} m before encountering one of these imperfections and being scattered with a consequent loss of kinetic energy. (The electron kinetic energy is converted into motional energy of the lattice, that is, into *heat*. This is the reason why all ordinary materials suffer a rise in temperature when they conduct an electric current.) Therefore, in order to gain 5 eV of kinetic energy in a distance of 10^{-8} m, an electron must be accelerated by a field of 5×10^8 V/m! This field is about 10^{10} times that which will cause current to flow in metallic crystals of Li, Na, K, etc. Diamond is therefore an extremely good insulator. Similar energy gaps in crystals such as quartz (SiO_2) and in plastics such as lucite make these materials good insulators.

STIMULATED EMISSION OF RADIATION

The probability that a transition between two states of an atom will take place within a certain specified time interval depends on the product of the wave functions describing those states and a quantity characteristic of the transition. Thus, from the quantum mechanical viewpoint, there is no distinction between the transition from state A to state B $(A \rightarrow B)$ and the transition $B \rightarrow A$ because the same product of wave functions and transition quantity is involved in the description of each transition. The two equivalent processes, excitation and deexcitation, are shown schematically in Fig. 11.26; in each case the photon has an energy $h\nu = \mathcal{E}_B - \mathcal{E}_A$.

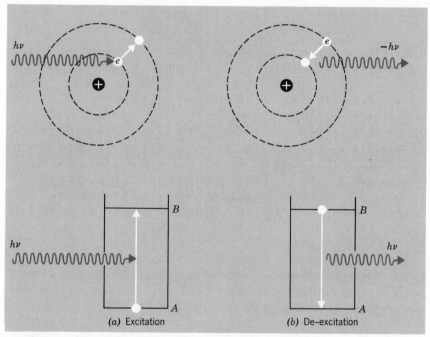

Fig. 11.26 (a) *An incident photon of energy* hν *excites an atom by raising an electron to a higher energy state. (b) Deexcitation occurs when the electron returns to its ground-state configuration and a photon of energy* hν *is emitted. According to the rules of quantum mechanics, the two processes are mathematically equivalent. The only difference between the two situations is that an energy* hν *is absorbed in excitation and an energy* hν *is emitted in deexcitation.*

It follows, therefore, that if we can arrange to have an atom in state B (the excited state) when a photon with an energy $h\nu = \mathcal{E}_B - \mathcal{E}_A$ is incident on that atom, then this photon will stimulate the deexcitation process to occur. (The photon cannot *excite* the atom because it is already excited, so it does the equivalent—it *deexcites* the atom.) This process is called *stimulated emission* and is illustrated schematically in Fig. 11.27.

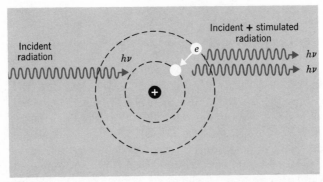

Fig. 11.27 *Stimulated emission of radiation. The incident photon of energy* h*ν finds the atom in an excited state and stimulates its decay. The two photons, each with energy* h*ν, proceed away from the atom in phase.*

The essential feature of the stimulated emission process that renders it both interesting and useful is that the incident photon and the stimulated photon proceed away from the atom *in phase*. That is, the photons travel in the same direction with their amplitudes oscillating together. The two photons therefore reinforce one another. If we had a sample of atoms, some fraction of which were in the same excited state, then a single incident photon could begin triggering the deexcitation of these atoms by stimulated emission. Each stimulated photon could, in turn, cause other atoms to emit photons and the entire system would radiate its excitation energy almost at once with a single bundle of photons all in phase. What have we gained in such a process? Since all of the excited atoms would eventually have radiated away their excitation energy by spontaneous emission, we have done no more by stimulating the deexcitation than would have happened anyway. The difference is that in the spontaneous emission process, the photons are radiated in random directions and not in phase. Stimulated emission produces the photons essentially simultaneously and in phase, thus concentrating the radiant energy in a narrow beam.

How is it possible to take advantage of stimulated emission to produce an intense beam of in-phase radiation? If the radiation is light, the device that accomplishes this is called a *laser*, an acronym for *l*ight *a*mplification by *s*timulated *e*mission of *r*adiation.

There are two main problems: First, how do we pump energy into the system of atoms so that a sufficient number of atoms are in the upper state? Second, how do we arrange for most of the photons to be emitted along the same direction?

If the high-energy state has a sharply defined energy (such as state B in Fig. 11.26), the pumping radiation must consist of photons with well-defined energy. A source of *white* light would not be suitable because such a source emits photons with a wide range of photon energies and so only a few of these can have the proper energy to be effective in exciting the atoms to the upper state. In 1960, Charles Townes and Arthur Shawlow of Columbia University called attention to an interesting property of ruby crystals that appeared to offer a solution to this problem. Ruby consists of aluminum oxide, a colorless substance, which contains a small amount of chromium as an impurity. The chromium impurity gives to ruby its characteristic red color. Figure 11.28 shows some of the energy states of the chromium atoms in ruby. Because there are broad bands at ε_2 and ε_3 (above the level that

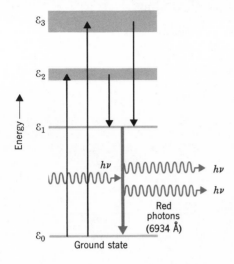

Fig. 11.28 *Some of the energy states of chromium atoms in a ruby crystal. The pumping radiation (upward arrows) excite the two energy bands, ε_2 and ε_3 which subsequently radiate to form the state ε_1 which exhibits laser action. The laser radiation (red arrow) consists of red photons ($\lambda = 6934 \ Å$)*

exhibits laser action), the white light from the source includes large numbers of photons whose energies fall within the range that permits the excitation of these energy bands. Each of the two bands radiate primarily to the state at ε_1. Hence, the laser transition is $\varepsilon_1 \to \varepsilon_0$ and the corresponding radiation is in the red part of the spectrum at 6934 Å.

The problem of directionality can be solved in the following way. A crystal of ruby is formed into a cylinder with the end surfaces accurately parallel (Fig. 11.29). One end is silvered to form a mirror while the other end is given only a partial coating of silver so that some of the radiation can escape from this end. The excitation is provided by a high-intensity lamp that spirals around the cylindrical crystal. As soon as one photon is produced in the spontaneous transition $\varepsilon_1 \to \varepsilon_0$, this triggers the laser action. Those photons that move parallel to the cylinder axis are reflected at the ends and again transverse the crystal, stimulating the emission of additional photons. A fraction of this radiation escapes through the partially-reflecting surface and constitutes the laser beam. Most of the spontaneously emitted photons are not emitted parallel to the axis; these photons are reflected in the crystal and eventually escape through the sides. These spontaneous photons do not contribute to the beam, but a sufficient number of photons *are* reflected back and forth to sustain the laser action.

Energy is continually pumped into the crystal by the light source and some fraction (usually very small) emerges as the laser beam; this radiation is in phase, has an almost pure frequency, and is highly directional. But in no sense is a laser a "source" of energy. In fact, only a very small fraction of the input energy appears in the beam. But *all* of this output energy appears in a tiny beam of small cross sectional area that is highly mono-chromatic. Some recently constructed lasers can produce bursts of radiation

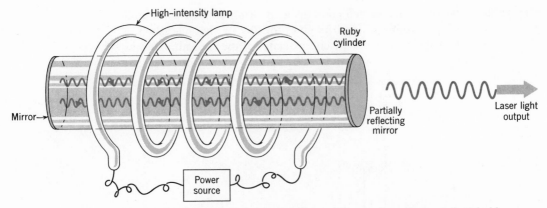

Fig. 11.29 *Schematic of a ruby laser system. The pumping radiation is furnished by a high-intensity source of white light. The stimulated photons are reflected back and forth between the parallel mirrors and build up the intensity of the radiation. The beam is formed by photons escaping through the partially reflecting surface.*

in which tens of joules of energy is released within 10^{-12} sec. This rate of energy output exceeds[2] 10^{13} watts (!) and can be delivered to areas smaller than 1 mm².

Hundreds of materials—solids, liquids, and gases—have beeen found to exhibit laser action. These lasers have rapidly found an extraordinary number of applications in basic research, technology, and medicine. Some of the more spectacular uses have been in eye operations where a laser has been found ideal for depositing just the right amount of energy to "weld" a detached retina onto the choroid surface that lies beneath it. Microholes can be drilled in hard substances by laser beams, and the welding of materials that resist other methods can be accomplished with these devices. Modulated laser beams can carry an incredible number of communications channels and it is clear that the impact on the communications industry of the development of such devices will be enormous. By reflecting a laser beam from a mirror placed on the moon we are now obtaining information regarding the fluctuations of the Earth-moon distance—information that will give important clues concerning the geophysics of the Earth and the structure of the moon.

11.9 *Superconductors*

CURRENTS THAT FLOW FOREVER

When an electric field is applied to a conductor, the free electrons are set into motion and thereby produce a current. Resistance to the flow of current in a metallic crystal is caused in part by collisions of the electrons with impurities or with points of imperfection in the crystal lattice structure. However, even in an ideal crystal that is both pure and perfect, so that these

[2] This is approximately 100 times the total electrical generating capacity of the U.S., but the laser pulse lasts for only $\sim 10^{-12}$ sec.

sources of resistance are absent, the electrons cannot flow unimpeded because the thermal vibrations of the atoms provide sites from which the electrons can be scattered and with which they can exchange energy. As the temperature is decreased, the thermal vibrations are lessened and the motion of the electrons is less violently affected. Thus, the resistance to current flow decreases as the temperature decreases.

In 1911 Kammerlingh Onnes discovered that lead has the remarkable property that at a temperature of 7.2°K the electrical resistance suddenly becomes zero—not just "very small" but zero! At temperatures of 7.2°K or lower, lead is a *superconductor*. In one experiment, a current of several hundred amperes was induced to flow in a highly refined sample of lead shaped into a ring, and this current was found to be still flowing, apparently undiminished, after a period of a year! The resistance of superconducting lead has been measured to be at least 10^{11} times smaller than the resistance of normal lead. There is every reason to believe that in pure samples of superconducting materials, the electrical resistance is indeed *zero*. Several elements and many alloys (over 1000 are known) have now been found to be superconductors at low temperatures.

The phenomenon of superconductivity is the result of macroscopic quantum effects. The basic idea of the theory of superconductivity is concerned with the fact that not all types of particles obey the exclusion principle. Electrons (and all other particles with spin $\frac{1}{2}\hbar$) obey the exclusion principle, but particles with integer spin $(0, \hbar, 2\hbar, \text{etc.})$ do not. In superconducting materials, the interaction of the conduction electrons with the vibrations of the atoms in the lattice overcomes the repulsive Coulomb force and results in a small net *attraction* between the electrons. Consequently, the electrons tend to group into *pairs* and a pair of electrons, with spins opposite, behaves as a particle with zero spin.

The net attraction between the electron pairs—the *pairing energy*—is very small and it does not require much agitation to break the pairs. Therefore, it is only at very low temperatures that the pairs can exist. Because they are not restricted by the exclusion principle, the electron pairs all tend to collect in the lowest possible energy state as the temperature is reduced. When the critical temperature is reached (7.2°K for the case of lead), all of the pairs are in the lowest state and all have the same wave function that extends throughout the material. None of the pairs can change its energy state, and therefore the electrons all flow together and there is no dissipation of energy and no electrical resistance.

Certain aspects of superconductors are still not well understood and are currently under investigation, but the crucial point in the explanation of the phenomenon is contained in the pairing of the electrons and we no longer consider superconductivity to be the great mystery that it once was.

Superconducting materials are beginning to be widely used in the construction of magnets for both research and technological applications. Electromagnets that produce strong magnetic fields are expensive to operate because of the substantial losses due to resistance effects in the windings. A conventional electromagnet that produces a field of 10^5 gauss (about the largest field that can be achieved with this type of magnet) requires an enormous amount of electrical power to maintain the field. Furthermore, such a magnet requires a cooling system that uses thousands of gallons of water per minute to prevent the windings from melting because of the generation of heat by resistance effects. Magnets are now being used in which the windings are made from various superconducting materials operated at temperatures below the critical temperature. Once the current is established in the windings of such a magnet, it continues to flow without resistance losses. Of course, no practical superconductor can be absolutely pure and so some energy losses do occur. But only very small amounts of input power are required to maintain the superconducting field.

Metallic alloys and compounds have been found to be more useful than pure elements in the construction of windings for superconducting magnets. A widely used material is Nb_3Sn, which allows the production of fields up to 88 kilogauss. By using V_3Ga, it is expected that fields as large as 500 kilogauss can be achieved.

If the resistance losses in the transport of electrical power could be eliminated or substantially reduced, enormous savings in cost would be realized. Therefore, the possibility of using superconducting materials for the construction of electrical transmission lines is of great economic importance. Perhaps within the near future we shall begin to replace the huge steel towers that now carry our electrical power with underground superconducting electrical lines.

Summary of Important Ideas

Rutherford's analysis of α-particle scattering experiments showed conclusively that most of the mass of an atom is concentrated in a tiny, positively-charged *nuclear* core.

In order to account for the lines in the hydrogen spectrum. Bohr found it necessary to postulate that each line corresponds to a transition between two allowed *discrete energy states* and that the angular momentum of the atom is limited to *discrete multiples of $\hbar$*. Bohr departed from classical electromagnetic theory by postulating that no radiation occurs except during the transition process.

According to the Bohr model, an integer number of de Broglie electron waves must exactly fit into every allowed electron orbit.

The specification of the quantum mechanical state of an electron in an atom requires *four* quantum numbers: n, l, m_l, and m_s, which specify, respectively, the (gross) *energy*, the *angular momentum,* the *component of the angular momentum* in a particular direction, and the *orientation of the spin vector* relative to the angular momentum vector.

The *Pauli exclusion principle* states that no two electrons in an atom can have exactly the same set of four quantum numbers. This principle accounts for the occurrence of *electron shells* in atoms.

The *ionic binding* of two (or more) atoms to form a molecule results when an electron is transferred from one atom to another so that attractive electrostatic forces bind the atoms together. The binding is *strong* when the removal or the addition of only one or two electrons leaves a *closed* electron shell.

ATOMS AND
MOLECULES

264

The *covalent binding* of atoms to form molecules results when two electrons are shared between the atoms.

The *exclusion principle* prevents more than two electrons ($m_s = \pm\frac{1}{2}$) from occupying a given energy state. In bulk material, these states are associated with the entire structure rather than with a single atom. Therefore, the atomic energy states are distributed over a certain energy range and become *energy bonds*.

In a *conductor*, a portion of the highest energy band that contains electrons is available for electrons that acquire additional energy. (That is, transitions *within* the band are possible; such transitions require very little energy.)

In an *insulator*, the highest energy band that contains electrons is *completely filled*. Therefore, no transitions *within* the band are allowed and excitations are possible only when an electron is carried into the next empty band; such excitations require considerably more energy than excitations in conductors.

When a photon stimulates the emission from an atom of a photon with the same frequency, the two photons propagate away from the atom *in phase* and reinforce one another. The operation of *lasers* is based on this fact.

The phenomenon of *superconductivity* is the result of *electron pairs* collecting in the lowest possible energy state and moving together without energy losses.

Questions

11.1

An *absorption spectrum* is one that results when "white" light (that is, light consisting of all frequencies) is passed through a substance. The absorption lines are then *dark* lines on a background of "white" light. What lines are found in the absorption spectrum of hydrogen?

11.2

A corollary to the exclusion principle is the principle of *indistinguishability* of elementary particles. This principle states, for example, that there is no way to distinguish any one electron from another electron. Contrast the situation in which one billiard ball collides with another billiard ball to that in which one electron collides with another electron. Can one measure the angle through which the *incident* object was scattered in both situations? (The billard balls are *numbered*, but what about the electrons?)

11.3

In what positions in the periodic table to you expect to find elements that have *low* photoelectric work functions? Compare your answer with the elements in Table 10.2.

11.4

Why are neutral atoms of lithium more chemically active than Li^+ ions?

11.5

Sodium and chlorine combine to form NaCl. Two chlorine atoms combine to form Cl_2. Why does sodium not form an Na_2 molecule?

11.6

Describe the way magnesium and chlorine combine to form a molecule. (Refer to Fig. 11.18 and decide how many electrons magnesium can contribute.)

11.7

Use diagrams similar to those in Fig. 11.14 and show schematically how the N_2 molecule is formed with three covalent bonds and how the O_2 molecule is formed with two covalent bonds.

11.8

Sketch the arrangement of the electrons in molecules of MgO and Na_2S.

11.9

Argue why it is not possible to form molecules consisting of two helium atoms (He_2). (Consider covalent bonding.)

11.10

Use the exclusion principle to show that three hydrogen atoms cannot be bound together by covalent bonding. (However, the H_3^+ *ion* can exist. Why?)

11.11

Do you expect the noble gases (in solid crystalline form) to be good electrical conductors? Explain.

Problems

11.1

An α particle of energy 5.3 MeV from a radioactive source of Po^{210} approaches a gold nucleus "head on." How close to the nucleus can the α particle penetrate before being stopped and deflected backward? (That is, at what distance will the electrostatic potential energy equal the initial kinetic energy of the incident α particle?) It was from such a calculation that Rutherford was able to show that nuclei are much smaller than atoms.

11.2

What is the longest wavelength photon that can induce a transition in a hydrogen atom in its ground state? When that atom deexcites, in what series will the radiation be?

11.3

What is the longest wavelength photon that can ionize a hydrogen atom in its ground state? How would you classify this photon—visible, infrared, or ultraviolet?

11.4

What frequency must a photon have in order to raise a hydrogen atom from its ground state to the state with $n = 4$? Is this a "visible" photon?

11.5

A beam of 12.5-eV electrons bombards a quantity of hydrogen gas and excites some of the atoms. Photons with what energies will be emitted? (Refer to Fig. 11.5 and Example 11.1)

11.6

Use the information in Fig. 11.5 and Example 11.1 and calculate the energies of the three lowest energy lines in the Lyman series.

11.7

Construct a diagram similar to Fig. 11.12 for the hypothetical situation in which electrons have spin $\frac{3}{2}\hbar$. What elements will be in the K shall? With what element will the L shell close? (What are the possible values of m_s in this case?)

11.8

The ionization energy of potassium is 4.3 eV and the electron affinity energy of chlorine is 3.7 eV. In KCl the separation between the K^+ and Cl^- ions is approximately 3 Å. What is the molecular binding energy of KCl? (Begin by computing $PE_E = ke^2/R$; consider the ions to be point charges.)

11.9

Sulfur crystals are pale yellow and transparent. Sulfur is one of the better insulators. From this information alone, estimate the magnitude of the energy gap between the conduction band and the highest filled band in sulfur crystals.

11.10

Consider an atom with states of the following energies: -13.2 eV (ground state), -11.1 eV, -10.6 eV, -9.8 eV. Only the state at -11.1 eV exhibits laser action. The state at -10.6 eV radiates primarily to the state at -11.1 eV. The state at -9.8 eV radiates primarily to the ground state. What wavelength radiation would you use to pump the laser? What is the wavelength of the laser radiation?

CHAPTER 12
NUCLEI AND PARTICLES

The development of nuclear physics has been intimately connected with the history of atomic structure theory and the emergence of the concepts of quantum mechanics. The first hint that there were unknown forces at work within the atom was the discovery of *radioactivity* in the 1890s. Rutherford used the α particles that are emitted in certain radioactive decay processes to probe the inner structure of atoms and, based on these findings, he developed the nuclear model of the atom. Bohr elaborated on this concept and formulated the first crude theory of atomic structure. The culmination of this work came in the 1920s with the development of modern quantum theory. During the 1930s, while the advances made in quantum theory were being consolidated, a heightened interest in the nucleus produced a series of discoveries that projected nuclear physics to the forefront of basic research activities. The discoveries of the neutron, artificial radioactivity, fission, and the mu meson, and the development of the first theories of β radioactivity and of nuclear structure—these were all products of the 1930s. The last 30 years has seen attention turn to studies of the details of nuclear structure and of the way in which nuclear and subnuclear particles interact at extremely high energies. Perhaps these investigations will lead eventually to a fundamental understanding of the way Nature behaves in the nuclear domain.

12.1 *Properties of Nuclei*

PROTONS AND NEUTRONS

Although Rutherford in 1911 had concluded that all atoms have tiny massive nuclear cores, the atomic nucleus appeared to play no role in the many chemical and optical effects in which the atomic electrons participated. The nucleus seemed to be only the inert positively-charged core that attracted the active electrons and held the atom together. In the 1930s the nucleus itself came under detailed scrutiny. Nuclei were broken apart and their components studied. It was found that the nuclei of all atoms consist of only two fundamental types of matter. The first of these to be identified and studied was the *proton*, the nucleus of the hydrogen atom, which has a positive charge equal in magnitude to that of the electron and a mass that is 1836 times that of the electron. In 1932, James Chadwick discovered a second type of particle in nuclei—the *neutron,* an object with a mass almost equal to that of the proton but without an electrical charge.

An electrically neutral atom of a chemical element has a certain definite number of electrons in its outer shells and exactly the same number of protons in the nucleus. This number is called the *atomic number* of the particular element and is denoted by Z. A hydrogen atom has one electron and one proton and so has $Z = 1$; a helium atom has two electrons and two protons and so has $Z = 2$; an oxygen atom has $Z = 8$; and so on.

Although the nucleus of a hydrogen atom contains one proton and the nucleus of a helium atom contains two protons, a helium atom is approximately *four* times as massive as a hydrogen atom. The reason is that a helium nucleus contains two neutrons in addition to the two protons. In fact, the nuclei of *all* atoms (except hydrogen) contain neutrons as well as protons. The total number of protons and neutrons in the nucleus of an atom is called the *mass number* and is denoted by A. The number of neutrons in the nucleus is $N = A - Z$.

Atomic masses are measured on a scale in which the mass of an atom of carbon (the nucleus of which has 6 protons and 6 neutrons) is exactly 12 units. These units are called *atomic mass units,* AMU. On this scale, the proton and the neutron each have a mass of approximately 1 AMU:

$$\text{Proton mass} = m_\text{p} = 1.007276 \text{ AMU} \tag{12.1}$$
$$\text{Neutron mass} = m_\text{n} = 1.008665 \text{ AMU}$$

Therefore, the mass of a nucleus with mass number A is approximately equal to A AMU.

Expressed in grams, the atomic mass unit is[1]

$$1 \text{ AMU} = 1.6605 \times 10^{-27} \text{ kg} \tag{12.2}$$

NUCLEAR SIZES

Rutherford's early experiments showed that nuclei must be very much smaller than atoms—atoms have dimensions of $\sim 10^{-10}$ m whereas nuclei have dimensions of $\sim 10^{-14}$ m (see Fig. 12.1). Later experiments, particularly those involving the scattering of neutrons and electrons by nuclei, have shown that the radius of a nucleus with mass number A is given approximately by

$$R \cong 1.4 \, A^{\frac{1}{3}} \times 10^{-15} \text{ m} \tag{12.3}$$

This expression for the nuclear radius has the following meaning. If we calculate the *volume* of a nucleus, we find

$$V = \tfrac{4}{3}\pi R^3 = \tfrac{4}{3}\pi(1.4 \, A^{\frac{1}{3}} \times 10^{-15} \text{ m})^3$$
$$\cong A \times 10^{-44} \text{ m}^3$$

That is, the nuclear volume is simply proportional to the *total number* of protons and neutrons in the nucleus. The addition of protons and neutrons

[1] The relationship between AMU and kg is not yet known with sufficient precision to allow the adoption of an atomic standard for mass as we have done for length and time; see Section 1.3.

NUCLEI AND
PARTICLES

270

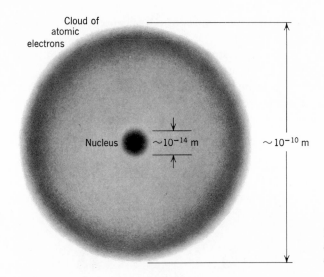

Cloud of
atomic
electrons

Nucleus $\sim 10^{-14}$ m

$\sim 10^{-10}$ m

Fig. 12.1 *Schematic comparison (not to scale) of atomic and nuclear sizes. The diameter of an atom is about 10,000 times that of a nucleus.*

to a nucleus to form a new element does not squeeze the particles any tighter; each proton and neutron occupies essentially the same volume independent of the number of these particles in the nucleus.

The extreme smallness of nuclei is difficult to comprehend. If we were to magnify an atom until it is the size of the Houston Astrodome, the nucleus located at midfield, would be no larger than a pea! The atomic electrons would also be pea-sized objects whizzing around somewhere in the upper and lower decks. Atoms, similar to galaxies, are mostly empty space. In fact, if we could compact all of the matter in the known Universe into one huge super-nucleus (without changing the density of nuclear matter), it would fit well within the limits of the solar system.

ISOTOPES

Nuclei of the same chemical element do not all have the same mass. Although most hydrogen atoms have nuclei that consist of a single proton, a small fraction of natural hydrogen atoms (about 0.015 percent) have one proton and one neutron in their nuclei. This "heavy hydrogen" is called *deuterium* (see Fig. 12.2). Another form of hydrogen atoms have nuclei with *two* neutrons; hydrogen with $A = 3$ is called *tritium*. The series of nuclei

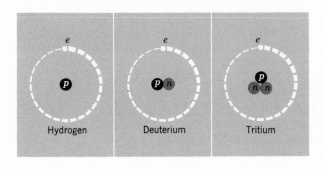

Hydrogen   Deuterium   Tritium

Fig. 12.2 *The three isotopes of hydrogen. Natural hydrogen consists predominantly of the $A = 1$ isotope. The $A = 3$ isotope (tritium) is unstable and undergoes radioactive decay. These simple schematic representations of atoms and nuclei are not realistic; atoms and nuclei are actually "fuzzy" objects (see Fig. 12.1)*

271

with a given value of Z but different values of A are called *isotopes* of the element.

Most elements have two or more stable isotopes; the average number is approximately 3, but tin has 10 isotopes. Different isotopes of a given element are distinguished by using the mass number as a superscript. Thus, the stable isotopes of helium are He^3 and He^4 (Fig. 12.3) and the stable isotopes of oxygen are O^{16}, O^{17}, and O^{18}. A list of some of the isotopes of the lightest elements is given in Table 12.1. The masses given are *atomic masses*.

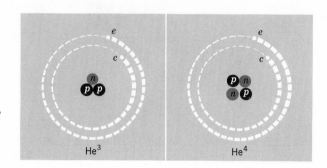

Fig. 12.3 *The stable isotopes of helium, He^3 and He^4. All other isotopes of helium are unstable. The abundance of He^3 in natural helium is only 1.5 parts per million.*

Table **12.1** *Properties of Some Light Elements*

Element	Z	A	Symbol	Atomic Mass (AMU)	Remarks†
Hydrogen	1	1	H^1	1.007 825	Stable (99.985%)
	1	2	H^2 or D^2 (deuterium)	2.014 102	Stable (0.015%)
	1	3	H^3 or T^3 (tritium)	3.016 050	β Radioactive
Helium	2	3	He^3	3.016 030	Stable (0.00015%)
	2	4	He^4	4.002 603	Stable (99.99985%)
	2	6	He^6	6.018 893	β Radioactive
Lithium	3	6	Li^6	6.015 125	Stable (7.52%)
	3	7	Li^7	7.016 004	Stable (92.48%)
	3	8	Li^8	8.022 487	β Radioactive
Beryllium	4	7	Be^7	7.016 929	Radioactive (e capture)
	4	8	Be^8	8.005 308	α Radioactive
	4	9	Be^9	9.012 186	Stable (100%)
	4	10	Be^{10}	10.013 534	β Radioactive

†The numbers in parentheses are the relative natural abundances of the isotopes.

NUCLEAR BINDING ENERGIES

At the very small distances that are characteristic of nuclear sizes, the strong nuclear force (Section 4.4) acts between pairs of nucleons to bind these particles into nuclei. The nuclear force is so strong (see Example 4.4) that an input of several MeV of energy is required to remove a nucleon from a nucleus. By using the Einstein mass-energy relation, $\mathcal{E} = mc^2$, we can discuss the *binding energy* of a nucleus in terms of its *mass*.

NUCLEI AND
PARTICLES

Consider, first, an atom of deuterium. The deuterium nucleus (called a *deuteron*) consists of one proton and one neutron; this is the simplest nucleus which depends on the nuclear force for its existence. The mass[2] of a deuterium atom (m_D) is slightly *less* than the combined masses of a hydrogen atom (m_H) and a free neutron (m_n):

$$m_H = 1.007\ 825\ \text{AMU} \qquad m_D = 2.014\ 102\ \text{AMU}$$
$$\underline{m_n = 1.008\ 665\ \text{AMU}}$$
$$m_H + m_n = 2.016\ 490\ \text{AMU}$$

Thus, for the mass difference, we have

$$(m_H + m_n) - m_D = 0.002\ 388\ \text{AMU}$$

This result is, in fact, quite general: all nuclei have masses that are *less* than the masses of the constituent protons and neutrons in the free state. The magnitude of this mass difference for a particular nucleus is indicative of the degree of *binding* of the protons and neutrons in that nucleus. According to the Einstein mass-energy relation, a *mass* difference corresponds to an *energy* difference. For the deuteron, this energy difference (the *binding energy*) is

$$\mathcal{E}_b = [(m_H + m_n) - m_D] \times c^2$$

In order to calculate $\mathcal{E}_b$ in MeV we need to know the value of c^2 in units of MeV/AMU. Using Eq. 12.2 and multiplying each side by c^2, we find

$$(1\ \text{AMU}) \times c^2 = (1.6605 \times 10^{-27}\ \text{kg}) \times (3 \times 10^8\ \text{m/sec})^2$$
$$\times \left(\frac{1\ \text{MeV}}{1.6022 \times 10^{-13}\ \text{J}}\right)$$
$$= 931.5\ \text{MeV}$$

Thus,

$$\boxed{c^2 = 931.5\ \text{MeV/AMU}} \qquad\qquad (12.4)$$

Physically, this equation means that the energy equivalent to 1 atomic mass unit is 931.5 MeV.

Returning to the calculation for the deuteron, we find

$$\mathcal{E}_b = (0.002\ 388\ \text{AMU}) \times (931.5\ \text{MeV/AMU}) = 2.224\ \text{MeV}$$

This result means that it is necessary to supply 2.224 MeV of energy to a deuteron in order to separate it into a free proton and a free neutron (or, equivalently, to separate a deuterium atom into a hydrogen atom and a free neutron). If more than 2.224 MeV is supplied, the excess energy will appear in the form of kinetic energy of the proton and the neutron.

The nuclear binding energy can also be interpreted in the following way. If, for example, a slowly moving neutron (that is, a neutron with negligible kinetic energy) is captured by the proton in a hydrogen atom to form a

---

[2]Isotopic masses are always listed in terms of the *atomic* mass instead of the *nuclear* mass, and we use the atomic masses here. Because the hydrogen atom and the deuterium atom each contain one electron, the difference in atomic masses is equal to the difference in nuclear masses.

deuteron, the initial mass-energy of the system, $(m_H + m_n)c^2$, is greater than the final mass-energy $m_D c^2$, and so the energy difference $\mathcal{E}_b$ is radiated in the form of a γ ray, as shown in Fig. 12.4.

Fig. 12.4 *Schematic representation of the capture of a slow neutron (i.e., a neutron with negligible kinetic energy) by a proton to form a deuteron. The deuteron binding energy is radiated in the form of a γ ray.*

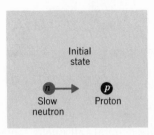
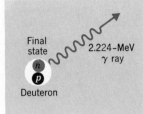

The deuteron has an exceptionally low binding energy; for most nuclei, the binding energy *per nucleon* is approximately 8 MeV. Thus, to separate a nucleus of Ne^{20} into 10 free protons and 10 free neutrons requires approximately $20 \times (8 \text{ MeV}) = 160$ MeV. Figure 12.5 shows the variation of $\mathcal{E}_b$ (in MeV per nucleon) with mass number. The decrease of $\mathcal{E}_b$ with mass number for $A \gtrsim 60$ has, as we will see, great significance in the phenomenon of *fission*.

12.2 *Radioactivity*

NUCLEAR DECAY PHENOMENA

At about the time that Thomson was investigating the properties of electrons, another discovery of great importance was made by the French physicist, Henri Becquerel (1852–1908). Certain naturally occurring minerals were found by Becquerel to emit radiations of a type that had not previously been observed. Within a few years, the emanations from *radioactive* substances were classified into three groups:

1. *Alpha rays*—massive, positively-charged objects.
2. *Beta rays*—negatively-charged objects of small mass.
3. *Gamma rays*—neutral rays with no detectable mass.

Detailed investigations of these radiations revealed that the beta rays are identical to electrons and that the alpha rays are nuclei of helium atoms. Gamma rays were found to have properties similar to light—the only difference being that the frequency of gamma rays is much higher than that of visible light.

Radioactivity is a *nuclear* phenomenon. Alpha and beta rays are emitted during the spontaneous disintegration of nuclei, and gamma rays result when the neutrons and protons within a nucleus spontaneously rearrange themselves (but without "disintegration").

The emission of an alpha ray (or α particle) by a nucleus necessarily changes both the atomic number and the mass number; that is, a new chemical element (a *daughter* element) is formed by the α decay of a *parent* nucleus. An α particle has $Z = 2$ and $A = 4$—it is just the nucleus of a *helium* atom. Therefore, when radium ($Z = 88, A = 226$) emits an α particle, radon

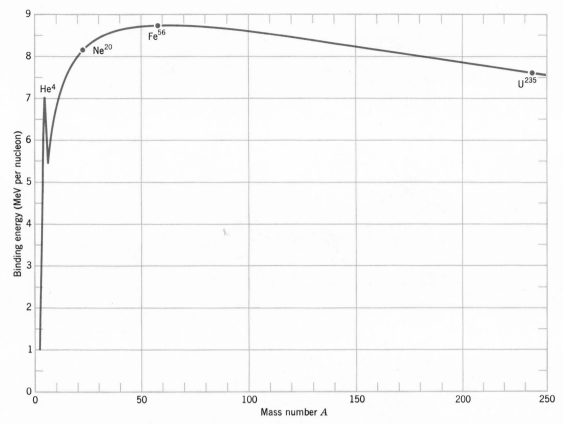

Fig. 12.5 *The binding energy per nucleon as a function of mass number. The point for He⁴ is far above those of the neighboring nuclei because the He⁴ is nucleus (the α particle) is an exceptionally tightly bound group of nucleons. The curve reaches a maximum in the vicinity of iron (Fe⁵⁶).*

($Z = 86$, $A = 222$) is formed, as indicated schematically in Fig. 12.6. Beta decay, on the other hand, does not involve the emission of a proton or neutron, so the mass number of the daughter nucleus is the same as that of the parent nucleus. But because the emitted particle carries a negative charge, the atomic number of the daughter nucleus is one unit *greater* than that of the parent. As shown in Fig. 12.6, when C^{14} ($Z = 6$, $A = 14$) decays by beta emission, N^{14} ($Z = 7$, $A = 14$) is formed. Beta decay is therefore equivalent to the transformation of one of the nuclear neutrons into a proton. The β particles emitted by nuclei are exactly the same as atomic *electrons*. Gamma rays are emitted when a nucleus makes a transition from a higher energy state to a lower energy state. This process, which leaves Z and A unaltered, is similar to an atomic transition, except that the energy involved is considerably greater.

HALF-LIFE

Since α and β emissions cause the parent nuclei to be transformed into different nuclear species, will not these disintegrations soon cause the complete depletion of the parent substance? Actually, this is not the case.

12.2
RADIO-
ACTIVITY

275

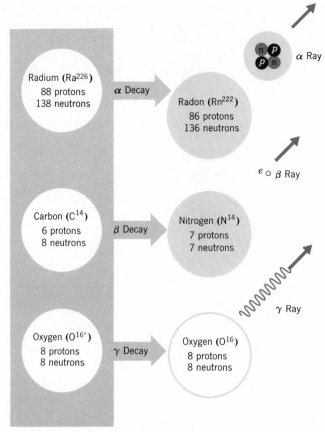

Fig. 12.6 *Examples of the three types of radioactive decay events. Alpha and beta decay involve nuclear disintegrations (i.e., changes in species) while gamma decay results from intranuclear rearrangements. The excited O^{16} nucleus that exists before γ decay takes place is indicated by O^{16*}.*

Table **12.2** *Some Radioactive Half-Lives*

Nucleus	Type of Decay	Half-Life
Thorium (Th^{232})	α	1.4×10^{10} y
Plutonium (Pu^{239})	α	100 y
Uranium (U^{229})	α	58 min
Carbon (C^{14})	β	5568 y
Cobalt (Co^{60})	β	5.3 y
Copper (Cu^{66})	β	5 min
Krypton (Kr^{94})	β	1.4 sec

Radioactive decay processes obey the following law: If we begin with an amount of a radioactive substance, then after a certain interval of time that is characteristic of the particular nucleus involved (called the *half-life* of the substance and denoted by $\tau_{1/2}$), *one-half* of the material will have disintegrated and one-half will remain. If we wait for another interval $\tau_{1/2}$, one-quarter of the original amount will remain. After each period of time $\tau_{1/2}$, there will remain one-half of the parent material that existed at the beginning of that time period. Radium-226, for example, has $\tau_{1/2} \cong 1600$ years; therefore, if we have 1 g of Ra^{226} to start with, after 1600 years we shall have $\frac{1}{2}$ g, after 3200 years we shall have $\frac{1}{4}$ g, after 4800 years we shall have $\frac{1}{8}$ g, and so forth (see Fig. 12.7). In quantum mechanical terms the statement of the radioactive decay law is that the probability for the decay of a given atom within a certain specified time interval is always the same regardless of how long the atom has existed. Thus, the probability for decay within a time interval $\tau_{1/2}$ is always exactly *one-half*.

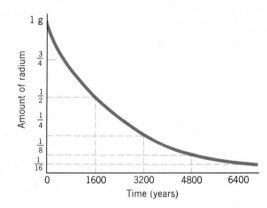

Fig. 12.7 *Radioactive decay curve of Ra^{226} ($\tau_{1/2} \cong 1600$ y). In each interval of 1600 years, the amount of radium decreases by one-half.*

β RADIOACTIVITY

Within a nucleus, a neutron is a stable particle just as is a proton. But, the mass of a *free* neutron is 1.008665 AMU, which is greater than that of a proton and an electron (that is, a hydrogen atom, $m_H = 1.007825$ AMU). Therefore, it is energetically possible for a free neutron to separate into a proton and an electron, and, in fact, neutrons do undergo this type of decay process with a half-life of 10.8 min.

The conversion of a neutron into a proton and an electron is the prototype of all nuclear β decay processes. Many nuclei (some of which occur naturally and others which must be produced artificially in the laboratory) are known to undergo β decay by the emission of electrons in exactly the same way that free neutrons decay.

Accompanying the electron in every β decay process there is emitted an additional particle called a *neutrino*. The neutrino has unusual properties—it has no mass and has no electrical charge but it carries linear momentum, angular momentum, and energy! Furthermore, the neutrino interacts only weakly with all forms of matter and so is extremely difficult to detect. Although the neutrino was postulated by Pauli in 1930 and was incorporated into a successful theory of β decay a few years later, it was not until 1953 that an experiment sufficiently sensitive to detect the neutrino was actually carried out.

All nuclear radioactive decay processes in which an electron is emitted can be considered to be the result of the β decay of a neutron *within* the parent nucleus. For example, consider the case of Li^8 (Fig. 12.9). The Li^8 nucleus consists of 3 protons and 5 neutrons and undergoes β decay with

Table **12.3** *Some Important β-Radioactive Nuclei*

Nucleus	Half-Life	Type of Decay	Maximum Electron Kinetic Energy (MeV)	Remarks
H^3 (tritium)	12.26 yr	β^-	0.0186	Used in nuclear fusion devices (H-bombs)
C^{14}	5730 yr	β^-	0.156	Used in archeological dating; also an important tracer in biochemical studies
Na^{22}	2.60 yr	β^+	0.54	Useful source of positrons
Na^{24}	15.0 hr	β^-	1.39	Used in medical diagnostics to follow the flow of sodium in the body
K^{40}	1.3×10^9 yr	β^-	0.0118	Used in archeological dating
Co^{60}	5.24 yr	β^-	0.31	Accompanying γ rays used in medical therapeutics and for radiation processing of plastics, food, etc.
Sr^{90}	28.8 yr	β^-	0.54	Important fission product (occurs in fallout from detonation of fission bombs)
I^{131}	8.05 days	β^-	0.61	Used in medical diagnostics and therapeutics, particularly in thyroid ailments

a half-life of 0.85 sec. The Li^8 decay process transforms one of the 5 neutrons into a proton so that the new nucleus has 4 protons and 4 neutrons—this new nucleus is Be^8.

POSITRON DECAY

Figure 12.10 shows the relative masses of the various nuclei with $A = 65$. For this mass number, only Cu^{65} is stable; Co^{65} has a mass greater than that of Ni^{65} and, therefore, undergoes β decay, forming Ni^{65}; furthermore, Ni^{65} has a mass greater than that of Cu^{65} and undergoes β decay, forming stable Cu^{65}. That is, two successive β decays transform a nucleus of Co^{65} into Cu^{65}. (The simultaneous emission of two electrons, which would transform Co^{65} directly into Cu^{65}, does not occur with a measurable rate.)

The $A = 65$ chart also shows that there are nuclei with $Z > 29$ that are more massive than Cu^{65}. These nuclei cannot undergo normal β decay because each has a charge *greater* than that of the adjacent, less massive

NUCLEI AND
PARTICLES

278

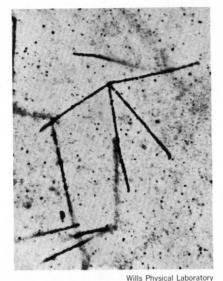

Fig. 12.8 *One method of rendering visible the path of a* single *nuclear particle is through the use of special photographic emulsions (called* nuclear emulsions). *This photomicrograph shows the tracks left by several α particles emitted in the radioactive decay of a single original thorium atom. First, thorium emits an α particle, leaving a radioactive atom; this atom emits an α particle, leaving another radioactive atom; and so on. The length of the longest track is approximately 0.03 mm.*

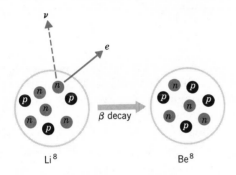

Fig. 12.9 *Schematic representation of the β decay of Li⁸. One of the Li⁸ neutrons is transformed into a proton and the new nucleus is Be⁸.*

nucleus. A new type of β decay process is possible for these nuclei which allows them to rid themselves of one unit of the excessive positive charge that they carry. Nuclei such as Ge^{65}, Ga^{65}, and Zn^{65} emit particles that are identical to electrons in every respect except that they have a *positive* charge. These particles are called *positrons* (β^+) and the decay process is called *positron decay*. We shall learn more about these particles when we discuss the properties of elementary particles later in this chapter. For now it suffices to note that there are two complementary nuclear β-decay processes, which we label β^- decay and β^+ decay; these processes can be represented in the following way:

$$\beta^- \text{ decay:} \quad n \longrightarrow p + e^- + \bar{\nu}_e \tag{12.5}$$

$$\beta^+ \text{ decay:} \quad p \longrightarrow n + e^+ + \nu_e \tag{12.6}$$

We understand, of course, that these processes take place *within* nuclei; the electron (or positron) and the neutrino are *created* at the instant of disintegration (they do *not* preexist in the nucleus) and are immediately ejected from the nucleus.

Positron-emitting nuclei do not occur naturally in our world—these nuclei

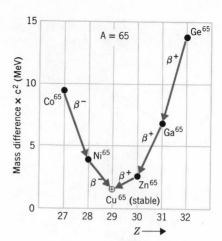

Fig. 12.10 *Nuclear β^- decay and positron (β^+) decay routes for nuclei with A = 65. There is a single stable nucleus with this mass number (Cu^{65}).*

must be produced artificially by means of nuclear reactions. The positron was discovered in 1932 and two years later the first example of artificial positron radioactivity was produced.

α DECAY

Certain unstable nuclei, primarily those with mass numbers above 200, spontaneously emit helium nuclei (α particles). The emission of an α particle by a nucleus decreases the original nuclear charge by 2 units and decreases the original mass number by 4 units. If a nucleus identified by (Z, A) has a mass greater than the sum of the masses of the nucleus $(Z - 2, A - 4)$ and a He^4 nucleus, the nucleus (Z, A) is unstable and can decay by the emission of an α particle.

Essentially all nuclei with $A \gtrsim 100$ are unstable with respect to breakup by the emission of α particles. But it is only for nuclei with $A \gtrsim 200$ that α decay is an important process. The heavier nuclei have α-decay half-lives sufficiently short that α-particle emission from a given sample occurs at a rate that permits observation, but the half-lives of the lighter nuclei tend to be so long that the α-decay process is unmeasurable, even though it is energetically allowed.

RADIOACTIVE DECAY CHAINS

The heaviest elements found in Nature are uranium (U, $Z = 92$), protactinium (Pa, $Z = 91$), and thorium (Th, $Z = 90$). All of the isotopes of these elements are radioactive but each element has at least one isotope with a sufficiently long half-life that the element still exists in Nature. For example, U^{238} has $\tau_{1/2} = 1.4 \times 10^9$ years. When these nuclei decay, they form new daughter elements that are also radioactive. Some of these nuclei are β radioactive and others emit α particles. A few can even decay by either α or β emission. A series of successive radioactive decays takes place that continues until a stable isotope of either lead (Pb, $Z = 82$) or bismuth (Bi, $Z = 83$) is formed. The stable isotopes of lead are Pb^{206}, Pb^{207}, and Pb^{208}; only Bi^{209} is stable. These four nuclei are the termination points for all of

Table **12.4** *Some Typical α-Radioactive Nuclei*

Nucleus	Half-Life	α-Particle Kinetic Energy (MeV)	Remarks
Ce^{142}	5×10^{15} yr	1.5	Lightest naturally occurring α-radioactive nucleus
Po^{210}	138 days	5.30	Much used source of α particles
Bi^{214}	19.7 min	5.51	Also undergoes β decay
Po^{218}	3.05 min	6.00	Formed by two successive α decays starting with Ra^{226}; also known as radium-A; used in original Rutherford scattering experiment
Ra^{226}	1620 yr	4.78	α particles from this source first identified as helium nuclei (Rutherford)

the radioactive decay chains that originate with the long-lived heavy elements. One such decay chain begins with U^{238} and ends with Pb^{206}; this series of α and β decays is shown in Fig. 12.11.

THE STABILITY OF NUCLEI

Figure 12.12 is a schematic chart of nuclei arranged according to proton number (Z) and neutron number (A–Z). There are more than 1600 known nuclei; some 330 nuclear isotopes have been found to occur naturally in the Earth (about 260 of which are stable) and almost 1300 have been produced artificially in the laboratory.

A number of interesting points appear on examination of the systematics of nuclear properties when the nuclei are arranged in this fashion:

1. The stable nuclei up to $Z \cong 20$ have approximately equal numbers of protons and neutrons ($Z \cong N$).

2. For $Z \gtrsim 20$, the stable nuclei tend to have an increasing preponderance of neutrons over protons; for example, uranium has $N/Z \cong 1.6$. The reason for this effect is easy to understand when we recall that the nuclear force has a *short* range whereas the Coulomb force has a *long* range. As more nucleons are added to form heavier nuclei, the average distance between nucleons becomes greater. Therefore, the long-range Coulomb repulsion becomes more effective relative to the short-range nuclear attraction and it becomes more and more difficult to add protons to a nucleus. For this reason it becomes energetically more favorable to add neutrons to heavy nuclei; consequently, N/Z increases with Z.

3. The stable nuclei are located along a narrow band of the $N - Z$ diagram, called the *valley of stability*. The nuclei on either side of the valley

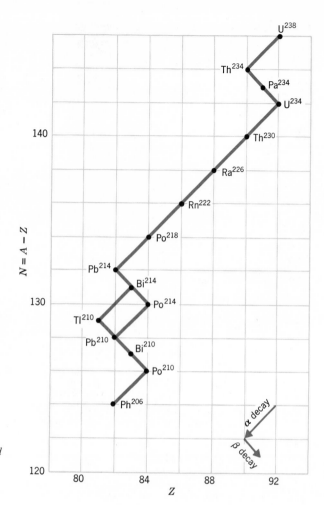

Fig. 12.11 *The radioactive decay chain that originates with U^{238} and ends with Pb^{206}. Notice that Bi^{214} can undergo either α or β decay and that after one additional decay, both branches lead to Pb^{210}. Each Bi^{214} nucleus has a certain probability for decay by β-emission and a certain probability for decay by α-particle emission.*

have larger masses and undergo radioactive decay (by either β^- or β^+ emission) in order to reach a stable condition. (Of course, for the heavier nuclei, α decay is possible, and some nuclei undergo spontaneous fission in an effort to reach stability.)

4. Nuclei *above* the valley of stability are *neutron rich* and undergo β^- decay. An increase in the distance from the stable valley causes: (a) the nuclei to become more massive (that is, less stable), (b) the β decay energy $\mathcal{E}_\beta$ to increase, and (c) the half-life to decrease. Sufficiently far from the valley, the neutron excess becomes very large and the instability increases to such a degree that the β^- decay process, which converts a neutron into a proton, is replaced by the direct emission of a neutron.

5. Nuclei *below* the valley of stability are *proton rich* and undergo β^+ decay. Similar to the case above the valley, nuclei sufficiently far below the valley will be short-lived emitters of high-energy positrons or will decay by proton emission.

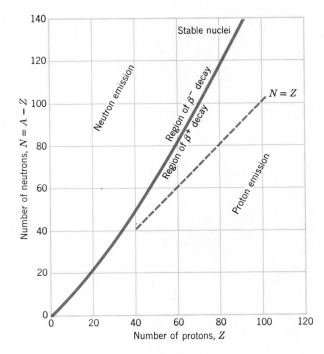

Fig. 12.12 *Schematic chart of nuclei. The narrow band is the valley of stability. Nuclei that lie above this valley decay by β⁻ or by neutron emission and those that lie below the valley decay by β⁺ or by proton emission. Nuclei with Z ≲ 20 have N ≅ Z, but for higher atomic numbers, N > Z.*

12.3 *Nuclear Reactions*

TRANSMUTATIONS OF NUCLEI

In 1919, Rutherford used α particles from a radioactive source to bombard nitrogen gas and found that occasionally an α particle would react with a nitrogen nucleus to produce a proton. The nuclear reaction that Rutherford observed (the first such to be discovered) was

$$N^{14} + He^4 \longrightarrow O^{17} + H^1$$

Usually, we write such reactions in a shorter notation as $N^{14}(\alpha, p)O^{17}$, where the first symbol is the target nucleus, the second is the bombarding particle (an α particle), the next is the outgoing particle (a proton), and the last is the residual nucleus. A cloud-chamber[3] photograph of a $N^{14}(\alpha, p)O^{17}$ reaction was first taken in 1925 by P. M. S. Blackett; this photograph is reproduced in Fig. 12.13.

In all nuclear reactions we must have a balance of protons and neutrons in the initial and final states. For the $N^{14}(\alpha, p)O^{17}$ reaction, we have

$$\left.\begin{array}{lccccc} & N^{14} & + & He^4 & \longrightarrow & O^{17} + H^1 \\ \text{No. protons:} & 7 & + & 2 & = & 8 + 1 \\ \text{No. neutrons:} & 7 & + & 2 & = & 9 + 0 \end{array}\right\}$$

Using this rule, we can always identify the fourth nucleus in a reaction if the other three nuclei are known.

[3] A *cloud chamber* is a device that renders visible the track of a nuclear particle by virtue of the condensation of water droplets on the ions left in the wake of the particle.

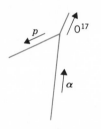

P. M. S. Blackett

Fig. 12.13 *Cloud-chamber photograph of a $N^{14}(\alpha, p)O^{17}$ reaction amidst the tracks of many α particles that do not induce reactions.*

Example **12.1**

B^{10} is bombarded with neutrons and α particles are observed to be emitted. What is the residual nucleus?

$$B^{10} + n \longrightarrow (?) + He^4$$

No. protons: $5 + 0 = Z + 2$

No. neutrons: $5 + 1 = N + 2$

Clearly, $Z = 3$ and $N = 4$, therefore, the residual nucleus is Li^7.

THE DISCOVERY OF FISSION

Shortly before his death in 1937, Lord Rutherford stated that "the outlook for gaining useful energy from the atoms by artificial processes of transformation does not look very promising." Although Rutherford's intuition in scientific matters was almost always infallible, within a few years, a series of scientific and technological advances had shown this particular view to be incorrect—incorrect, in fact, to an astonishing degree.

In 1939, the German radio-chemist, Otto Hahn, in collaboration with Fritz

NUCLEI AND
PARTICLES

284

Strassman, bombarded uranium with neutrons and performed very careful chemical tests on the resulting radioactive material. They found that among the products of neutron absorption by uranium there was radioactive barium ($Z = 56$) an element much less massive than the original uranium. How could such a light element be formed from uranium? The mystery was soon resolved by Lise Meitner and Otto Frisch, German physicists working then as refugees in Sweden, who suggested that neutron absorption by uranium produced a breakup (or *fission*) of the nucleus into two light fragments:

$$U(Z = 92) + n \longrightarrow Ba(Z = 56) + Kr(Z = 36)$$

This was a startling new type of nuclear reaction. Instead of exchanging only a few nucleons between the incident particle and the target nucleus, as in an (α, p) reaction, this discovery showed that it was possible to split a nucleus into two massive parts.

THE DYNAMICS OF FISSION

As shown in Fig. 12.5, the binding energy of a heavy nucleus ($A \cong 240$) is approximately 7.5 MeV per nucleon. If such a nucleus were separated into two parts, each with $A \cong 120$, the binding energy would be *increased* to approximately 8.5 MeV per nucleon. This change in binding energy per nucleon with mass number means that a heavy nucleus can break up into two light fragments with the release of a substantial amount of energy. This breakup of a heavy nucleus is somewhat analogous to the splitting apart of a vibrating drop of liquid, as shown in Fig. 12.14. Even though it is energetically favorable for a heavy nucleus (such as U^{235}) to split into two parts, this process is inhibited by the strong attractive nuclear forces. The nucleus may become extended in an effort to fission (as in the left-hand sequence in Fig. 12.14), but it will usually return to and vibrate around its equilibrium shape. The probability of the occurrence of *spontaneous fission* is extremely small, and therefore the corresponding half-life is extremely long ($\sim 10^{17}$ years for U^{235}). If, however, some additional energy is supplied to the nucleus in the form of the binding energy of a captured neutron, this increase in energy may produce a large nuclear deformation which will be sufficient to permit the relatively easy separation of the nucleus into two fragments; thus, fission can occur (as in the right-hand sequence in Fig. 12.14).

The amount of energy released in a fission event is approximately 200 MeV per event. This is truly an enormous amount of energy. If 1 kg of U^{235} undergoes fission, approximately 8×10^{13} joules of energy is released. This amount of energy is sufficient to raise the temperature of 200,000,000 gallons of water from room temperature to the boiling point. The original atomic bombs of 1945 contained about 2 kg of fissionable material. A photograph of the result of an atomic bomb test is shown in Fig. 12.15.

CHAIN REACTIONS

The fact that the fission process releases several neutrons (usually 2 or 3) makes possible a series or chain of neutron-induced fission events that is self-sustaining. If one neutron from a fission event triggers the fission of

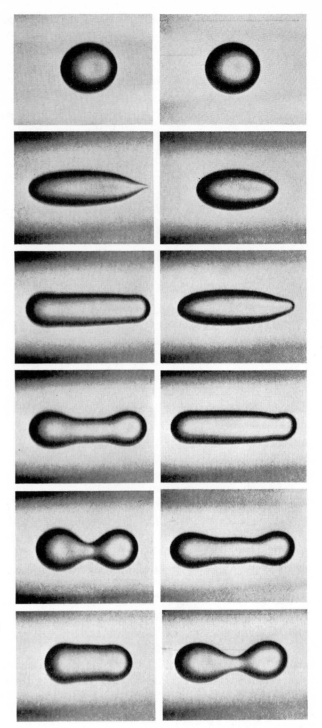

Fig. 12.14 *Photographs from a motion picture film showing the variation in shapes of an ordinary drop of water suspended in oil when a deformation is induced by a voltage applied across the oil. In the left-hand sequence, the drop returns to its initial spherical shape without undergoing fission. In the right-hand sequence, the initial deformation is sufficiently large that the drop fissions. In 1939, Niels Bohr and John Wheeler proposed a liquid-drop model of nuclear fission that was successful in explaining the general features of the fission process.*

NUCLEI AND
PARTICLES

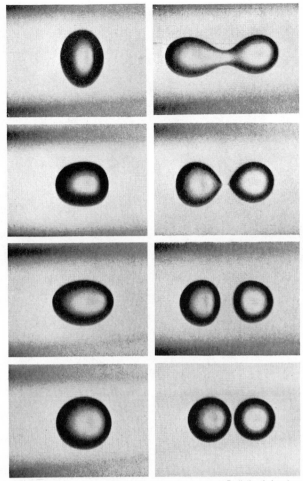

Fig. 12.14 (*Cont.*)

another nucleus and one neutron from this event triggers another fission, etc., this series of fission events will sustain itself and will constitute a *chain reaction.* By controlling the environment of the fissioning nuclei it is possible to maintain a condition in which each fission event contributes, on the average, one and only one neutron that triggers another event. In this way, the rate of energy generation (the *power*) is maintained at a constant level. The controlled fission chain reaction (Fig. 12.16) is the principle of the *nuclear reactor,* now widely used in the commercial generation of electricity.

It is also possible to bring together in a small volume a sufficient amount of fissionable material so that fewer of the fission neutrons escape the system and therefore more than a single neutron from each event can trigger a new event. Figure 12.17 shows a series of fission reactions in which each event contributes *two* neutrons toward the next set of events. The rapid multiplication of the number of fissioning nuclei in this uncontrolled situation leads to the explosive release of the fission-generated energy—this is the principle of the atomic bomb (which is, of course, actually a *nuclear* bomb). Because such a huge amount of energy is released in a very brief time in a localized space, the destructive effect of a nuclear weapon is

Fig. 12.15 *Underwater detonation of an atomic bomb at Bikini atoll in the Pacific.*

incredibly large. In 1945 the only two nuclear weapons to be used in warfare devastated the Japanese cities of Hiroshima and Nagasaki. About 100,000 persons (approximately one-quarter of the population of the two cities) were killed in these blasts. The delayed effects of the radiation exposure are still being studied.

FUSION

Energy will be released from any group of particles that can be rearranged into a system that has a greater binding energy. Fission, of course, is one example of such a process—the total binding energy of two nuclei such as barium and krypton is greater than the total binding energy of uranium,

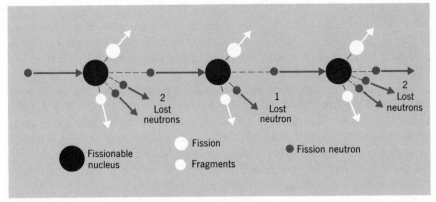

Fig. 12.16 *A controlled fission chain reaction in which one neutron from each fission event triggers another event. One or two neutrons from each fission event escape the system and are "lost."*

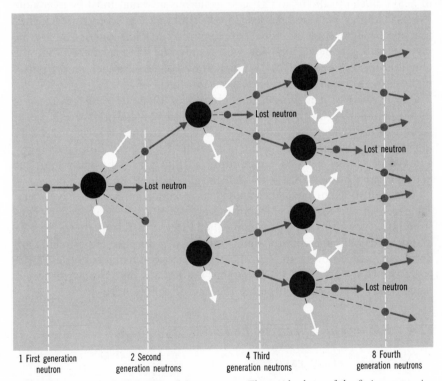

Fig. 12.17 *An uncontrolled series of fission events. The rapid release of the fission energy in such a system leads to an explosion.*

and so the fission process releases energy. The problem of extracting energy from nuclei can also be approached from the low-mass side of the maximum in the binding energy curve. If we combine two light nuclei to form a tightly bound medium-A nucleus, energy will be released. This process is called *fusion*. For example, if two Ne^{20} nuclei (binding energy per nucleon $\cong 8$ MeV; see Fig. 12.6) are combined to form a Ca^{40} nucleus (binding en-

12.3
NUCLEAR
REACTIONS

ergy per nucleon $\cong 8.5$ MeV), there would be a total energy release of 40×0.5 MeV $= 20$ MeV. The difficulty in this particular case, of course, is that a great force would be required to overcome the Coulomb repulsion and to bring the neon nuclei into sufficiently close proximity so that the capture process would take place.

The effect of the Coulomb repulsion will be reduced if we use nuclei with small Z. If we bring two deuterons together, two reactions are possible: $D^2 + D^2 \rightarrow He^3 + n$ and $D^2 + D^2 \rightarrow T^3 + H^1$. The energy released in each of these reactions is approximately 4 MeV. Thus, the average energy released in each D + D reaction is ~ 1 MeV per nucleon involved; this amount of energy release is comparable with that for the fission of a heavy element (200 MeV for 236 nucleons in the case of $U^{235} + n$).

Since it is relatively easy to separate deuterium from normal hydrogen, there is a vast supply of deuterium available to us in the form of *water*, particularly in the oceans. How can we make use of this enormous reservoir of energy? Coulomb repulsion, which works to our advantage in the splitting of heavy nuclei in the fission process, is an obstacle that must be overcome if fusion energy is to be released. In order for the deuterons to react, they must be brought close together. This requires that their kinetic energies be high. Heating a deuterium gas will supply the kinetic energy, but the temperature required is $\sim 10^7 \, °$K! Reactions that require such an extremely high temperature are called *thermonuclear reactions*.

One method of achieving a temperature of 10 million degrees or so is by the detonation of an atomic (fission) device. In the brief fraction of a second during which the blast takes place, the temperature is sufficiently high to ignite thermonuclear reactions, which then release additional energy and maintain the elevated temperature so that all of the thermonuclear material can "burn." This is, in fact, the principle of the H-bomb.

Although uncontrolled thermonuclear reaction processes have been achieved (in the form of H-bombs), we have not yet succeeded in constructing a device in which the controlled release of fusion energy can be maintained for longer than a small fraction of a second. Experiments are now being carried out with sophisticated devices in several countries (especially the U.S., U.K., and U.S.S.R.) and the hope is that a practical fusion reactor may be constructed before the end of this century.

12.4 *Biological Effects of Radiation*

TYPES OF RADIATION EXPOSURE

Every person on Earth is continually exposed to various kinds of radiation from many different sources. Ordinarily, these radiations do us no particular harm. But even the most familiar of radiations—solar radiation—can do damage to the skin or eyes if the exposure is too great. Infrared radiation (from a heating element) or microwave radiation (from a microwave cooking oven) can also cause serious burns if used carelessly. Usually, however, when we speak of "radiation damage," we refer to the effects of high energy radiation, such as X rays from television sets or medical devices and γ rays from radioactive materials. X rays and γ rays are sufficiently energetic that they can penetrate to any point in the human body, whereas ultraviolet radiation is absorbed completely in the skin. Therefore, X and γ radiation can affect the internal organs and nervous system, whereas UV radiation can at most produce a severe sunburn.

The high-energy radiation to which the general public is exposed is almost exclusively in the form of X rays or γ rays. Radiation workers, on the other hand, sometimes come into contact with materials that emit α and β particles. All of these radiations can produce biological damage by virtue of their ionizing action in living tissue. Because an α particle is much more massive than an electron, the ionization produced by an α particle is much more localized than electron-induced ionization. A biological system can cope with distributed ionization much more readily than it can with concentrated radiation. As a result, the biological damage produced by a 5-MeV α particle is about 10 times as severe as that produced by a 5-MeV electron or by a 5-MeV γ ray (which produces energetic electrons in the ionization process).

Radiation exposure is measured in terms of a unit called the *rad*, which stands for *radiation absorbed dose*. If 1 kg of material absorbs 0.01 J of radiation energy, the dose is said to be 1 rad:

$$1 \text{ rad} = 0.01 \text{ J/kg} \tag{12.7}$$

A dose of 10 rad of X or γ radiation is approximately equivalent (in terms of biological damage) to a dose of 1 rad of α particles.

RADIATION DAMAGE

The effects of radiation can be classified as *somatic* (effect on the individual exposed) or *genetic* (effect on the offspring of the individual exposed). The somatic effects at various radiation dose levels are summarized in Table 12.5. Delayed somatic effects include the increased susceptibility to leukemia, bone cancer, and eye cataracts.

Genetic effects of human exposure to radiation are much more subtle than somatic effects, and we still know relatively little about the way in which radiation-damaged genes produce mutations. We do not even know the radiation level at which mutation production becomes important. For this reason, the permissible exposure to radiation for the general public has been set very much lower than the level at which somatic effects become evident. The maximum allowable whole-body exposure of an individual during one year has been set at 0.5 rad of γ radiation (or a correspondingly lower figure if the exposure is to α particles). The average per capita annual dose is shown for various sources of radiation in Table 12.6. This is the *average* dose and so the exposure of a particular individual to power-plant radiation, for example, could be considerably higher. Generally speaking, however, the radiation from natural sources (cosmic rays and natural radioactivity in the earth, building walls, etc.) and that from medical instruments far outweighs, at the present time, the radiation from any other source. Because of the increasing number of nuclear power plants, the radiation dose attributable to this source will continue to increase. But even by the year 2000, it has been estimated that the nuclear power dose level will still be, on the *average*, a tiny fraction of the dose from natural sources. Nevertheless, the utmost precautions must be taken in choosing sites for nuclear power plants, in the operation of the reactors, and in the disposal of the radioactive wastes.

Table **12.5** *Somatic Effects of Radiation Exposure*

γ-Ray Whole Body Dose (rad)	Effects	Remarks
0–25	None detectable	
25–100	Some changes in blood but no great discomfort, mild nausea.	Some damage to bone marrow, lymph nodes, and spleen.
100–300	Blood changes, vomiting, fatigue, generally poor feeling.	Complete recovery expected; antibiotic treatment.
300–600	Above effects plus infection, hemorrhaging, temporary sterility.	Treatment involves blood transfusions and antibiotics; severe cases may require bone marrow transplants. Expected recovery about 50 percent at 500 rad.
>600	Above effects plus damage to central nervous system.	Death inevitable if dose >800 rad.

Table **12.6** *Average Radiation Exposure of an Individual in the U.S.*

Source	Average Per Capita Dose (rad/yr)†
Natural (cosmic rays, natural radioactivity)	0.130
Medical X rays	0.090
Weapons test fallout	0.005
Nuclear power	<0.00001
	0.225

†Whole-body dose of γ radiation.

These are probably the most challenging problems that face the power industry and the Atomic Energy Commission today.

MEDICAL USES OF RADIATION

X rays have long been used in the medical profession for diagnostic work, particularly with regard to bones and teeth. More recently, radioactive isotopes have been employed in specialized diagnostics and therapeutics. For example, iodine is selectively absorbed by the thyroid gland. The radioactive isotope I^{131} has a half-life of 8 days (see Table 12.3). Therefore, if a small amount of I^{131} is ingested by a patient, the iodine will have ample

NUCLEI AND PARTICLES

opportunity to find its way to the thyroid gland but it will not live sufficiently long to constitute any kind of long-term radiation hazard. By locating the absorbed I^{131} with radiation detection instruments, it is possible to determine, for example, whether a portion of the gland is cancerous. In addition, continued I^{131} treatments can arrest or destroy the cancerous growth. More than a dozen radioisotopes are currently used internally in selective diagnostics and therapeutics. Further, the γ rays from external sources of Co^{60} are routinely used to provide massive doses of radiation in the treatment of various types of cancers.

12.5 *Elementary Particles*

PARTICLES AND ANTIPARTICLES

While he was investigating cloud-chamber tracks of cosmic-ray particles in 1932, Carl D. Anderson (1905–   ) observed a track that appeared to be due to an electron. But this track curved the "wrong way" in the magnetic field in which the cloud chamber was located, indicating that the particle carried an electrical charge *opposite* to that of an electron. This was the first observation of a positively-charged electron (a *positron*).

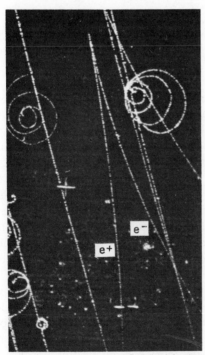

Fig. 12.18 *Bubble-chamber photograph of the creation of a positron-electron pair. An energetic γ ray enters the chamber from above and interacts with one of the (hydrogen) nuclei in the chamber to produce the pair. The chamber is located in a magnetic field, and so the tracks of the two particles curve in different directions.*

Soon after Anderson's discovery, it was established that positrons can actually be *created* by the interaction of energetic photons (γ rays) with matter. This creation process, however, always produces a *positron-electron pair* (Fig. 12.18), and therefore does not violate the general principle of charge conservation. In the creation of a positron-electron pair, electromagnetic energy is converted into mass; in order to create two electron masses, the photon energy must be at least $2m_ec^2 = 1.02$ MeV.

Once a positron is created, it interacts via electromagnetic forces with the atomic electrons in its vicinity, eventually losing essentially all of its kinetic energy. As the positron drifts with very low velocity it can encounter and coalesce with an electron. The two particles then *annihilate* one another and the mass-energy of the pair appears in the form of two photons with a total energy of $2m_ec^2$ (Fig. 12.19).

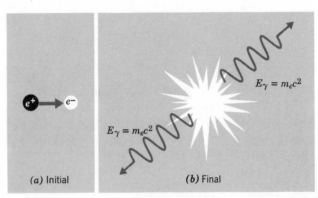

Fig. 12.19 *The annihilation of a slowly moving positron in an encounter with an electron produces two photons each with energy $\mathcal{E}_\gamma = m_ec^2$ (annihilation radiation). The photons leave the annihilation site "back-to-back" in order to conserve momentum.*

Electrons and positrons are said to be *antiparticles* of one another. The positron is the antiparticle of the electron and *vice versa*, but since the electron is the natural member of the pair in our world, we usually refer to the electron as the "particle" and to the positron as the "antiparticle."

All elementary particles have antiparticle partners. (The neutral pion $\pi°$ and the photon are in a special category—each of these particles is its *own* antiparticle.) A particle and its antiparticle have exactly the same mass, the same spin quantum number, and, if they are unstable, the particle and the antiparticle decay in the same way with the same half-life. However, the members of a particle-antiparticle pair have *opposite* electrical properties. Thus, the electron carries a negative charge and the positron carries a positive charge.

After the positron was discovered, it was natural to wonder whether *antiprotons* and *antineutrons* exist. As is the case for the positron, an antiproton can be produced only along with its antiparticle, a proton; similarly, an antineutron can be produced only along with a neutron. It requires an energy $2m_ec^2 = 1.02$ MeV to produce an electron-positron pair and; by the same token, it requires an energy $2m_pc^2 = 1876$ MeV to produce a proton-antiproton pair and an energy $2m_nc^2 = 1879$ MeV to produce a neutron-antineutron pair. The concentration of such huge amounts of energy in a single elementary particle that initiates the creation event can be achieved only in the largest accelerators, and such accelerators were not available until the 1950s. In 1955, however, a group working with the 6-GeV accelerator that had recently been constructed at the University of California was successful in producing and identifying antiprotons (symbol: $\bar{p}$); in the following year, the antineutron ($\bar{n}$) was discovered.

Electrons and positrons interact via the electromagnetic force and when an electron-positron annihilation event takes place, the products are the quanta of the electromagnetic field—photons. Nucleons interact primarily via the strong nuclear force and when a proton-antiproton or neutron-antineutron annihilation event takes place the products are *pions,* the quanta of the nuclear force field. Figure 12.20 shows the annihilation of an antiproton by a proton in a bubble chamber; in this event, 8 charged pions (and probably several neutral pions which leave no tracks) are produced.

Brookhaven National Laboratory

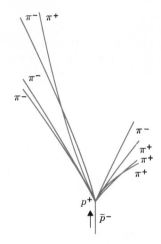

Fig. 12.20 *Annihilation of an antiproton with a proton in a hydrogen bubble chamber. Eight charged pions (and probably some neutral pions that do not leave tracks in the chamber) are produced.*

Antiprotons are *stable* particles, but a *free* antineutron, if it does not undergo annihilation, will eventually decay into an antiproton and a positron. Because the half-life of the antineutron is so long (presumably, the same as that of the neutron—10.8 min), annihilation will take place before decay occurs; the spontaneous decay of an antineutron has never been observed.

The most complex form of antimatter that has yet been produced and identified is *antihelium* ($2\bar{p} + 2\bar{n}$). Conceivably, *antiatoms,* consisting of antiprotons, antineutrons, and positrons, could be produced; but because they would annihilate immediately on contact with ordinary matter, no such complete antiatom has yet been identified.

STRONGLY INTERACTING PARTICLES

The types of elementary particles that interact via the strong nuclear force include pions, nucleons, and a new group of particles, called *hyperons,* that are short lived and have masses greater than the nuclear mass.

Pions are reasonably well understood because they are primarily responsible for propagating the strong interaction. As pointed out in Section 6.5, Yukawa realized the necessity for the existence of a particle of intermediate mass that would propagate the attractive force between nucleons. Although

Yukawa's prediction was made in 1935, the pion was not discovered until a 1947 experiment revealed the presence of these particles in cosmic rays. In the following year, pions were first produced artificially in an accelerator and since that time beams of pions have been available for use in detailed studies of their properties and interactions.

There are two types of charged pions, π^+ and π^- (which are the antiparticles of one another), and a neutral pion, π^0 (which is its own antiparticle). The charged pions are slightly more massive than the neutral pion: $m_{\pi^+} = m_{\pi^-} = 273\, m_e$ and $m_{\pi^0} = 264\, m_e$.

Pions can be produced in collisions in which an energetic nucleon or photon is incident on a nucleon; for example,

$$p + p \longrightarrow p + n + \pi^+$$
$$n + p \longrightarrow n + p + \pi^0$$
$$\gamma + p \longrightarrow p + \pi^0$$

The first pion decays to be observed took place in photographic emulsions exposed to cosmic rays (see Fig. 12.21). When a pion comes to rest in an emulsion, it emits a muon and a neutrino:

$$\pi^+ \longrightarrow \mu^+ + \nu_\mu$$
$$\pi^- \longrightarrow \mu^- + \bar{\nu}_\mu$$

The neutral pion, on the other hand, decays predominantly into a pair of γ rays:

$$\pi^0 \longrightarrow \gamma + \gamma$$

When an elementary particle decays, it always does so predominantly through the strongest (and therefore the *fastest*) interaction that is available to the particle. If it is allowed, decay via the strong interaction always dominates. When a proton is supplied with 300 MeV of energy, it forms an "excited state" in much the same way that a hydrogen atom forms its first excited state when it is supplied with 10.2 eV of energy. An excited hydrogen atom returns to its ground state by the emission of a quantum of the electromagnetic field that binds the atom together, namely, a *photon*. An excited proton returns to its ground state by the emission of a quantum of the strong interaction field, namely, a *pion*. This process is therefore a *strong* decay, and measurements have shown that the half-life for the decay of an excited proton is $\sim 10^{-23}$ sec. *This time is typical of all strong interaction processes.*

Although pions interact strongly with nucleons and are created in strong interaction processes, pions cannot *decay* via the strong interaction because there are no less massive particles that interact strongly. Therefore, pion decay must involve either the electromagnetic or the weak interaction. Charged pions cannot undergo a purely electromagnetic decay because the electric charge must be carried off by some kind of *particle*. The decay of charged pions is therefore restricted to the weak interaction, and the products of π^+ and π^- decays are muons and neutrinos—weakly interacting particles. Since decays via the weak interaction are slow processes, we expect charged pions to be relatively long-lived particles, and, indeed, the half-life is $\sim 10^{-8}$ sec. Neutral pions decay by gamma rays through the stronger electromagnetic interaction and have a half-life $\sim 10^{-16}$ sec. (Remember, these

Fig. 12.21 *Decay of a pion in a photographic emulsion exposed to cosmic rays. The pion comes to rest in the emulsion (at the bottom of the picture). In the decay process, a muon (μ) is emitted (along with a neutrino which leaves no track in the emulsion). The muon eventually comes to rest in the emulsion (at the top of the picture) and decays with the emission of an electron.*

Courtesy of C. F. Powell

times are to be compared with the strong-interaction time of 10^{-23} sec; therefore, a decay time of 10^{-8} sec represents a very slow decay.)

The first elementary particles to be found with masses greater than that of the proton were the *lambda particles*. Although these particles (Λ^0 and $\overline{\Lambda^0}$) are electrically neutral, they are readily identified by the V-shaped tracks that the charged decay products leave in emulsions or bubble chambers. Figure 12.22 shows a bubble-chamber photograph of a proton-antiproton collision that results in the production of a $\Lambda^0 - \overline{\Lambda^0}$ pair. Each lambda particle travels only a short distance in the chamber before undergoing decay:

$$\Lambda^0 \longrightarrow \pi^- + p; \qquad \overline{\Lambda^0} \longrightarrow \pi^+ + \bar{p}$$

Following the discovery of the lambda particles, several additional heavy particles were found; these particles bear the labels Σ (sigma particles), Ξ (xi particles), and Ω (omega particles). Particles in this group are collectively called *hyperons*. Hyperons are strongly interacting particles and all undergo

12.5
ELEMENTARY
PARTICLES

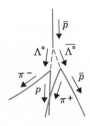

Lawrence Radiation Laboratory

Fig. 12.22 *Production of a $\Lambda^0 - \bar{\Lambda}$ pair by a $p - \bar{p}$ collision in a bubble chamber. The Λ^0 decays into a proton and a π^- meson, and the $\bar{\Lambda}^0$ decays into an antiproton and a π^+ meson.*

decays that lead to nucleons. But these decays take place via the slow weak interaction and so the half-lives are all long compared to the typical strong-interaction time of 10^{-23} sec (Table 12.7).

What are we to make of this collection of massive short-lived particles and their antiparticle partners? No one yet knows the fundamental significance of so large a number of "elementary" particles; indeed, there are many *additional* particles that exist for times shorter than the hyperons. In recent years various schemes for classifying these particles have led to an understanding of some of the interrelations among them, but it is apparent that we have not yet discovered the key that will allow us to penetrate to the heart of the puzzle of the strongly interacting elementary particles.

Table **12.7** *Hyperons*

Particle	Mass-Energy (MeV)	Half-Life (sec)	Anti-Particle
Λ^0	1115.5	1.7×10^{-10}	$\overline{\Lambda^0}$
Σ^+	1189.5	5.6×10^{-11}	$\overline{\Sigma^+}$
Σ^0	1192.5	$<10^{-14}$	$\overline{\Sigma^0}$
Σ^-	1197.4	1.1×10^{-10}	$\overline{\Sigma^-}$
Ξ^0	1314.9	2.0×10^{-10}	$\overline{\Xi^0}$
Ξ^-	1321.3	1.2×10^{-10}	$\overline{\Xi^-}$
Ω^-	1672	7.6×10^{-11}	$\overline{\Omega^-}$

LEPTONS AND THE WEAK INTERACTION

The group of weakly interacting particles (called *leptons*) consists of electrons, neutrinos, muons, and their antiparticles. Some of the processes that are governed by the weak interaction are nuclear β decay, the decay of charged pions, and the decay of muons:

NUCLEI AND
PARTICLES

298

β^- decay: $\quad n \longrightarrow p + e^- + \bar{\nu}_e$ ⎱
β^+ decay: $\quad p \longrightarrow n + e^+ + \nu_e$ ⎰

π^- decay: $\quad \pi^- \longrightarrow \mu^- + \bar{\nu}_\mu$ ⎱
π^+ decay: $\quad \pi^+ \longrightarrow \mu^+ + \nu_\mu$ ⎰

μ^- decay: $\quad \mu^- \longrightarrow e^- + \bar{\nu}_\mu + \nu_e$ ⎱
μ^+ decay: $\quad \mu^+ \longrightarrow e^+ + \nu_\mu + \bar{\nu}_e$ ⎰

In these processes we have indicated the neutrinos associated with electrons by a subscript e and the neutrinos associated with muons by a subscript μ; also, *anti*neutrinos have been distinguished by a bar over the symbol. That is, we have four different neutrinos: ν_e, $\bar{\nu}_e$, ν_μ, and $\bar{\nu}_\mu$. One of the more remarkable results of recent research in elementary particle physics has been the demonstration that these four types of neutrinos are all different and distinguishable. Thus, β^- decay *always* produces a $\bar{\nu}_e$ neutrino and *never* a ν_e neutrino nor one of the muon neutrinos. This fact has had a profound influence on the theory of weak interactions and we are now beginning to make progress in a fundamental understanding of these particles.

Summary of Important Ideas

Nucleons are bound together in nuclei by the strong nuclear force. The mass of any nucleus is *less* than the mass of the number of free protons and free neutrons that make up that nucleus; this difference in mass-energy is the total *binding energy* of the nucleus.

Not all groups of nucleons constitute stable nuclei; if there is a less energetic arrangement that is available to the nucleons (that is, if a configuration of smaller mass is possible), then a *radioactive decay* process will occur, which will transform the original nucleus into a nucleus of smaller mass. Radioactive decay involves one of the following possibilities: emission of an electron, emission of a positron, or emission of an α particle.

The stable nuclei with $A \lesssim 40$ contain approximately *equal* numbers of protons and neutrons ($Z \cong N$). For heavier nuclei, the neutron number increases more rapidly than the proton number.

When a nucleon or a nucleus is given a high velocity and is directed toward other nuclei, nuclear *reactions* can take place in which nuclear particles are emitted and new nuclei are formed.

Heavy nuclei (such as uranium) can absorb a neutron and undergo *fission* by splitting into two fragments of roughly equal mass. Each fission event releases approximately 200 MeV of energy. Energy is also released when two light nuclei combine to form a heavier nucleus; this process is called *fusion*.

All elementary particles have antiparticle partners (e^- and e^+; p and $\bar{p}$; etc.) Some particles (π° and photons) are their *own* antiparticles. The properties of a particle and its antiparticle are the same except that the electrical properties are *opposite*.

Pions are the quanta of the strong interaction and can be produced copiously in nucleon-nucleon collisions. Pions decay into muons and neutrinos.

The weakly-interacting particles (*leptons*) consist of e^-, μ^-, ν_e, ν_μ, and their antiparticles. The four types of neutrinos are all different and distinguishable.

Questions

12.1

The dominant stable isotope of oxygen is O^{16} ($Z = 8$, $N = 8$). The isotope O^{14} is radioactive. By what kind of process do you expect O^{14} to decay?

12.2

Why are there no pairs of *stable* nuclei with the same value of A but with Z differing by one unit?

12.3

If the nuclei (Z, A) and $(Z + 2, A)$ are both stable, what general statements can be made concerning the nucleus $(Z + 1, A)$? (Is this nucleus radioactive? If so, what type of decay does it undergo?)

12.4

The thorium isotope Th^{232} decays by α emission. What is the daughter nucleus formed in this decay? (Refer to Fig. 11.8 to determine Z.)

12.5

Neutron capture by a stable target nucleus rarely leads to positron radioactivity. Why?

12.6

What are the residual nuclei when a (p, α) reaction takes place with the following target nuclei: Be^9, B^{11}, O^{18}, and F^{19}?

12.7

A deuteron is captured by a B^{10} nucleus and a γ ray is emitted. What nucleus has been formed?

12.8

List some stable targets and incident particles that could be used to produce nuclear reactions that yield $N^{13} + n$ in the final state.

12.9

List some of the reactions that can take place when Be^9 is bombarded with protons.

12.10

When U^{235} ($Z = 92$) absorbs a slow neutron, it undergoes fission and releases 2 or 3 neutrons. List 3 or 4 possible pairs of fission-product nuclei that could be formed in such a process.

12.11

Fission reactors produce substantial amounts of radioactivity. Explain why this constitutes a certain hazard. Would *fusion* reactors suffer from this same defect?

12.12

An electron and a positron can bind together into an "atomic" system called *positronium.* What is "antipositronium?"

12.13

A beam of high energy γ rays strikes a target of He^3. Write down some of the possible photoproduction reactions that produce pions. (Consider reactions with the individual nucleons in the nucleus.)

Problems

12.1

What is the approximate radius of the Al^{27} nucleus?

12.2

What is the approximate *density* of nuclear matter? (Argue that all nuclei have approximately the same density.)

12.3

The atomic masses of naturally occurring elements are sometimes quite different from an integer number of atomic mass units because the element is actually a mixture of isotopes. Copper, for example, consists of two isotopes, Cu^{63} (69.1%) and Cu^{65} (30.9%). What do you expect the atomic mass of natural copper to be (approximately)? Compare your result with that given in Fig. 11.8.

12.4

What is the binding energy per nucleon of (a) He^3, (b) Li^6, and (c) Li^7? (Use the masses given in Table 12.1.)

12.5

If two Li^6 nuclei were brought together, what nucleus would be formed and how much energy would be released? (See Table 12.1 for the mass of Li^6.)

12.6

Use Fig. 12.5 and estimate the amount of energy that would be released if 20 protons and 20 neutrons were brought together to form Ca^{40}.

12.7

The mass of U^{238} is 238.0508 AMU. What fraction of the total mass-energy of U^{238} is its *binding energy?*

12.8

A sample of β-radioactive material is placed near a Geiger counter (a detector of β rays). The detector is found to count at a rate of 640 per sec. Eight hours later, the detector counts at a rate of 40 per sec. What is the half-life of the material?

12.9

An important method of determining the age of archeological items is by *radioactive carbon dating*. Radioactive C^{14} is produced at a uniform rate in the atmosphere by the action of cosmic rays. This C^{14} finds its way into living systems and reaches an equilibrium concentration of about 10^{-6} percent compared to normal, stable C^{12}. When the organism dies, C^{14} ceases to be taken up. Therefore, after the death of the organism, the C^{14} concentration decreases with time according to the radioactive decay law with $\tau_{\frac{1}{2}} = 5568$ years. An archeologist working a *dig* finds an ancient firepit containing some crude pots and bits of partially consumed firewood. In the laboratory he determines that the wood contains only 12.5 percent of the amount of C^{14} that a living sample would contain. What date does he place on the artifacts discovered in the dig?

12.10

The mass of Fe^{56} ($Z = 26$) is 55.934 936 AMU and the mass of Co^{56} ($Z = 27$) is 55.939 847 AMU. Which of these nuclei is stable and which decays radioactively into the other?

12.11

What is the energy of electrons emitted in the β decay of tritium? (Use the masses given in Table 12.1.)

12.12

A slow neutron is captured by Li^7 and a single γ ray is emitted. What is the energy of the γ ray? (Use the masses given in Table 12.1.)

12.13

What would be the total amount of mass-energy relased if an antihydrogen atom annihilated with an ordinary hydrogen atom? (Neglect the atomic binding energy.) What would be the products of the annihilation?

12.14

What is the amount of mass-energy released as kinetic energy in the decay $\Sigma^0 \rightarrow p + \pi^0$?

CHAPTER 13
ASTROPHYSICS
AND COSMOLOGY

For thousands of years Man has regarded the heavens with wonder and mystery—and no less so now than in the past. For in spite of the fact that modern science has made tremendous advances in understanding the macroscopic and microscopic world around us, our progress in answering the ancient questions concerning the Universe has been slow and stumbling. How did the stars originate—or have they existed "forever?" Why is there such a variety of different types of stars? Are there sources of energy in stars and galaxies that we have not yet discovered? How has the Universe evolved? What is its future? Although we have been able to give partial answers to these questions, each step forward always seems to open new areas of the unknown.

One of the great difficulties in studying astronomical phenomena is that we cannot perform controlled experiments on the objects of our study. We can only observe those events that are taking place naturally. This handicap has been alleviated somewhat by the construction of huge telescopes that gather light from galaxies billions of light years away, but we are still forced to accept for study only those events and objects that Nature provides.

Because of the severity of the limitations placed on the astrophysicist and the cosmologist, perhaps the greatest source of amazement is not the extraordinary diversity of astronomical happenings that we can see taking place, but the fact that we can analyze these events and, from them, draw conclusions regarding the course of stellar and galactic history over billions of years.

13.1 *Nuclear Reactions in Stars*

THE FORMATION OF STARS

Most of the matter in the Universe is in the form of hydrogen and helium. Wherever it can be measured or estimated—in the Sun, in major planets, and in stars—the abundance of helium is found to be about 10 percent of the hydrogen abundance by number of atoms (or about 30 percent by mass). Only a small fraction of this hydrogen and helium is contained in stars—the remainder is distributed throughout interstellar (and intergalactic) space.

In our Galaxy the average density of interstellar matter is about 1 atom per cm³. Because of the random motion of the atoms, the density in a particular region of interstellar space can become sufficiently high and the gravitational forces sufficiently large that a condensation of the matter will result. If the original gas cloud has a large mass (perhaps a thousand times the Sun's mass) the condensing gas can break up into a large number of small gas clouds, each of which can continue to condense under the influence of the local gravitational field. During this process some of the gas condenses into *dust grains*. These small clouds of gas and dust condense into individual stars, and, thus, a cluster of new stars is formed.

There is ample evidence to support the assertion that many stars are formed in clusters by condensation from clouds of gas and dust. The *Pleiades* (Fig. 13.1) is one of the most famous examples of an open cluster[1] of stars

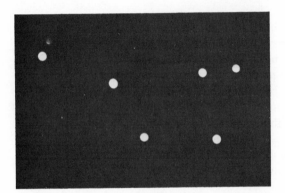

Yerkes Observatory

Fig. 13.1 *Two photographs of the Pleiades cluster. The short exposure on the left shows only the six naked-eye stars, whereas the long exposure on the right reveals many other stars as well as regions of nebulosity caused by the reflection of star light from residual dust clouds that were left after the stars condensed.*

presumably formed by cloud condensation. This cluster contains dozens of stars in a close grouping, six of which are visible to the naked eye as a small, dipper-shaped cluster in the constellation *Taurus*. Too many stars are concentrated within too small a region of space for the clustering to be due to any cause except condensation from a common source of material. The stars of the Pleiades were surely formed together. Furthermore, a long-exposure photograph (right-hand photograph of Fig. 13.1) reveals several prominent regions of nebulous luminosity in the Pleiades. This nebulosity is caused by star light reflected from dust clouds that remain after the condensation of the stars.

Some stars condense from almost pure hydrogen and helium, but other stars contain in addition the heavy-element material ejected from erupting stars (*novae*) and exploding stars (*supernovae*). For example, about 2 percent of the mass of the Sun is in the form of heavy elements.

ASTROPHYSICS
AND
COSMOLOGY

[1] An *open cluster* is an irregular and diffuse collection of stars, as distinct from the regular and concentrated collection of a large number of stars that constitutes a *globular cluster* (see Fig. 4.2).

As a star condenses from a cloud of gas and dust, gravitational potential energy is released. Some of this energy is radiated away and the rest is converted into kinetic energy of the condensing atoms—that is, the *temperature* of the star is increased. This gravitational energy was long believed to be the exclusive source of energy in stars. But in the 1920s it was realized that the amount of gravitational energy possessed by a star is insufficient to account for the vast quantity of energy radiated by a typical star (such as the Sun) throughout its lifetime of billions of years. Attention was then turned to thermonuclear reactions, initiated by gravitational heating, which could account for this enormous outpouring of energy.

When a star condenses from interstellar material, the first thermonuclear reaction to take place involves only hydrogen. A capture reaction takes place in which two protons form a deuteron and a positron and a neutrino are emitted:

$$H^1 + H^1 \longrightarrow D^2 + \beta^+ + \nu_e \tag{13.1}$$

When a condensing star has reached the stage at which the density in the central region is $\sim 10^5 \, kg/m^3$ and the temperature is $\sim 10^{7} °K$, the stellar protons have sufficient thermal energies for the $p + p$ capture reaction to begin to take place. Once deuterium is formed in this reaction, there rapidly follow two additional reactions which result in the production of helium:

$$D^2 + H^1 \longrightarrow He^3 + \gamma \quad \text{or} \quad D^2(p, \gamma)He^3 \tag{13.2}$$

followed by a reaction involving two He^3 nuclei:

$$He^3 + He^3 \longrightarrow 2H^1 + He^4 \quad \text{or} \quad He^3(He^3, 2p)He^4 \tag{13.3}$$

The net result of this series of reactions (called the *proton-proton* or *p-p chain*) is the conversion of four hydrogen atoms into one helium atom, as shown schematically in Fig. 13.2. The total amount of energy released in this series of reactions is 26.73 MeV. The γ rays and positrons that are produced in these reactions are absorbed by the gas in and surrounding the thermonuclear core and therefore contribute to the heating of the star. The neutrinos, on the other hand, because of their weak interaction with matter, escape from the star and carry away energy. Taking into account this energy loss, the average energy released in the star by each set of *p-p* chain reactions is approximately 26.3 MeV or about 6.5 MeV per nucleon. Each gram of hydrogen that is converted into helium releases approximately 6×10^{11} J of energy. In the Sun, the *p-p* reactions convert hydrogen at a rate of about 6×10^{11} kg/sec.

In general, the conditions required for thermonuclear reactions to occur are found only in the central region of a star. Energy released in these reactions is radiated as photons from the core to the surrounding material. These photons exert a *radiation pressure* (see Section 8.5) on the outer layers of the star's gas. In the equilibrium situation, the inward gravitational force exerted on any small volume of stellar material is just equal to the outward force caused by radiation. A star does not continue to contract after the thermonuclear reactions in the core begin to produce sufficient radiation to balance the inward gravitational force.

13.1
NUCLEAR
REACTIONS
IN STARS

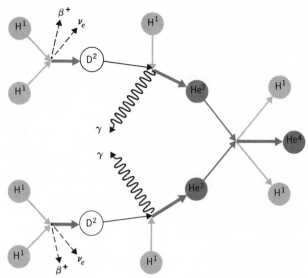

Fig. 13.2 *Schematic representation of the proton-proton chain which converts four hydrogen atoms into one helium atom. Six protons are involved in the sequence, but two are returned to the medium in the final reaction.*

THE CARBON-NITROGEN CYCLE

If a star contains some carbon, there is another series of nuclear reactions that can convert hydrogen into helium with the release of energy. In these reactions, carbon serves as a nuclear catalyst and is returned unconsumed, to participate in additional reactions. The reaction sequence is

$$\left.\begin{array}{l} \text{C}^{12}(p, \gamma)\text{N}^{13} \\ \text{N}^{13} \longrightarrow \text{C}^{13} + \beta^+ + \nu_e \\ \text{C}^{13}(p, \gamma)\text{N}^{14} \\ \text{N}^{14}(p, \gamma)\text{O}^{15} \\ \text{O}^{15} \longrightarrow \text{N}^{15} + \beta^+ + \nu_e \\ \text{N}^{15}(p, \alpha)\text{C}^{12} \end{array}\right\} \tag{13.4}$$

Thus, three protons are captured in a series of (p, γ) reactions and β decays; when a fourth proton is absorbed, an α particle is emitted, reforming a C^{12} nucleus. The net result is the same as that of the *p-p* chain, namely, the conversion of four hydrogen atoms into one helium atom with the release of about 6.5 MeV of energy per hydrogen atom consumed. Because this series of reactions proceeds by using and then reforming carbon and nitrogen, it is called the *CN cycle.* The cyclic nature of the reactions is illustrated in Fig. 13.3.

In a newly condensed star, the initial period of thermonuclear energy generation depends entirely on the proton-proton chain of reactions. But in stars that contain carbon, the CN cycle can compete with the *p-p* chain. The relative amounts of energy generated by the two processes in a given star depend on the temperature of the star's core. For temperatures below about $2 \times 10^{7\circ}$K, the *p-p* chain dominates, but when $T \gtrsim 2 \times 10^{7\circ}$K, the CN cycle is the primary source of energy. Thus, the more massive and brighter (and hotter) stars, such as the blue-white star *Sirius,* derive their

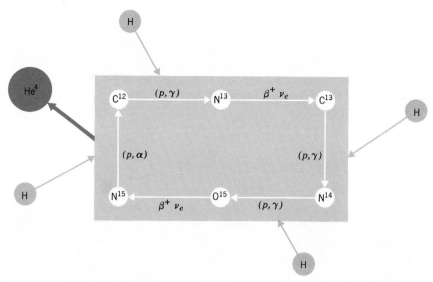

Fig. 13.3 *The carbon-nitrogen (CN) cycle of reactions that converts hydrogen into helium—four protons enter the cycle and one helium nucleus emerges. This reaction cycle was first proposed as a source of stellar energy by H. A. Bethe in 1937; Bethe was awarded the 1967 Nobel Prize in physics for his analysis of energy generation processes in stars.*

energy from the CN cycle. The Sun's primary source of energy is the *p-p* chain.

THE SYNTHESIS OF HEAVY ELEMENTS

When the hydrogen in the core of a star has been converted into helium by the *p-p* chain of reactions, the next stage in the synthesis of elements and the generation of energy takes place by the formation of C^{12} from three α particles in the two-step process, $He^4 + He^4 \rightarrow Be^8$ followed by $Be^8 + He^4 \rightarrow C^{12}$. Even though the Be^8 nucleus has a lifetime of only about 10^{-15} sec for breakup into two α particles, if the temperature and the density in the core of the star are sufficiently high ($\sim 10^{8\circ}$K and $\sim 10^9$ kg/m^3, respectively), some of the Be^8 nuclei will capture an α particle and form C^{12}. Carbon is then available for use in the CN cycle or further "helium burning" can proceed by such reactions as $C^{12}(\alpha, \gamma)O^{16}$, $O^{16}(\alpha, \gamma)Ne^{20}$, $Ne^{20}(\alpha, \gamma)Mg^{24}$, etc.

The further formation of heavy elements by thermonuclear reactions with hydrogen and helium in stars is inhibited by the Coulomb repulsion between charged particles. As the atomic number of the capturing nucleus increases, it becomes more and more difficult to force a proton or an α particle through the Coulomb barrier to form a heavier nucleus. Even in the cores of extremely hot stars, the thermal energies of the particles are insufficient to produce high-Z elements. However, *neutron* capture reactions are not impeded by Coulomb effects, and these reactions readily take place. It is now generally believed that neutron capture reactions in stars are responsible for the synthesis of all the heavy elements.

Neutrons can be produced in stars by reactions such as $C^{13}(\alpha, n)O^{16}$ and $Ne^{21}(\alpha, n)Mg^{24}$. Successive neutron capture reactions, together with the β

decay of radioactive species, can produce heavy elements. Neutron capture reactions occur in red giant stars, as well as in novae and supernovae.

If neutrons are produced rapidly enough, the density of neutrons present in a star can become sufficiently large to permit the capture of neutrons by radioactive isotopes before they undergo β decay. In fact, the radioactive elements above lead ($Z = 82$) are all formed by a series of neutron captures by radioactive nuclei.

13.2 *Radio Astronomy*

RADIO GALAXIES

In 1931, K. G. Jansky, of the Bell Telephone Laboratories, discovered that there are radio waves emanating from the Milky Way. These radio waves are emitted continuously and over a broad frequency spectrum. The first instrument designed specifically to receive these cosmic radio waves—a *radio telescope*—was built in 1936. There are now hundreds of radio telescopes in operation around the world. A typical steerable "dish" type

Fig. 13.4 *The 210-ft diameter radio telescope at Parkes, Australia. This instrument has been used for much of the radio mapping of the southern sky. In the foreground is a smaller, movable radio telescope which is used in conjunction with the large instrument for interference measurements.*

ASTROPHYSICS
AND
COSMOLOGY

antenna is shown in Fig. 13.4. These instruments have detected radio waves emitted by many different types of astronomical objects—the Sun, the moon, some planets, certain stars and gas clouds in our Galaxy, and other galaxies.

In the late 1940s, thousands of discrete radio sources, each occupying a small region in the sky, were discovered and catalogued. One of the most intense radio sources, discovered in 1948, is located in the constellation *Cygnus* and is known as *Cygnus A*. This was the first extragalactic source to be identified (1951) with an optically visible astronomical object.

More than 200 discrete radio sources have now been identified with optical galaxies. The "normal" galactic radio sources emit radio energy at rates $\sim 10^{31}$ J/sec, a small fraction of the energy output at visible wavelengths. (For example, the Sun radiates $\sim 10^{26}$ J/sec at visible wavelengths and our Galaxy contains $\sim 10^{11}$ stars. Therefore, the energy radiated by our Galaxy at visible wavelengths is $\sim 10^{37}$ J/sec.) Some "peculiar" galaxies, on the other hand, emit radio energy at rates of 10^{33}–10^{37} J/sec; Cygnus A, for example, emits $\sim 10^{37}$ J/sec of radio energy, many times the energy output of this galaxy at visible wavelengths. Only about a half dozen other radio galaxies are known to be as energetic as Cygnus A. The reason for the enormous release of radio energy from these peculiar galaxies is not known but it is probably associated with the acceleration of charged particles in the galactic magnetic fields.

QUASARS

In the early 1960s, a discovery of great importance was made when several radio sources, for which the positions had been determined with sufficient accuracy, were identified with certain visual objects that have an unusual blue color. It has been concluded that these objects are not simply visible radiogalaxies; the reasoning is based on two important observations. First, the photographic images are sharp and star-like, not "fuzzy" as is the case for galaxies that are too distant to be resolved into individual stars. Second, the emitted radiations show variations with time that have periods of the order of a day. Because the radiation from an object cannot change significantly in a time less than the time required for light to cross the object, these unusual objects can be no larger than about one light-day (that is, $\sim 3 \times 10^{14}$ m, or about 200 astronomical units), extremely small compared to galactic sizes. These peculiar star-like objects are called *quasi-stellar objects* or *quasars*.

Further observations have shown that the spectra of quasars exhibit extremely large red shifts. If these red shifts are interpreted in terms of the general expansion of the Universe (see Section 1.10), then some of the quasars are among the most rapidly moving and most distant objects in the Universe. For a single star to lie in the outer regions of the Universe and to have the brightness observed for the quasars means that these objects are pouring out fantastic amounts of energy—$\sim 10^{39}$–10^{40} J/sec, about 10^{12}–10^{13} times the output of the Sun. At this rate, a quasar would radiate an amount of energy in just one month equal to the entire mass-energy of the Sun! In order to account for such lavish expenditures of energy, quasars must have masses that are $\sim 10^{9}$ times the Sun's mass.

Although it is not absolutely certain, it seems most probable that quasars are both extremely remote and extremely massive. How do such objects originate? What is their history? In what way are they related to galactic phenomena? Are quasars isolated, star-like objects or do they lie within galaxies and outshine all of the other galactic stars put together? At present we do not know the answers to any of these questions.

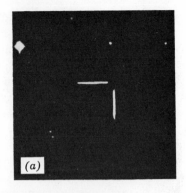

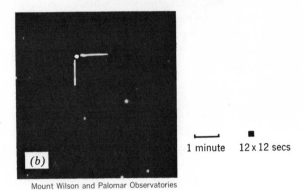

Fig. 13.5 *The intersections of the white lines indicate the positions of discrete radio sources. In (a) the field in the vicinity of the radio source 3C49 appears empty and no identification with an optical object can be made. In (b) the position of the radio source 3C309.1 corresponds almost exactly with a star-like object—a quasar.*

(a)

(b)

├────────┤   ■
1 minute   12 x 12 secs

Mount Wilson and Palomar Observatories

RADIO EMISSIONS FROM INTERSTELLAR HYDROGEN

The presence of atomic hydrogen in space can be detected by virtue of the distinctive radiation (which has a wavelength of 21 cm) that is emitted when the proton spin angular momentum vector and the electron spin angular momentum vector spontaneously change their relative orientation. Radio telescopes tuned especially to this radiation have been used to map the interstellar hydrogen in our Galaxy. Not only have regions of hydrogen concentration been detected, but the *velocities* of the atoms in these regions (along the lines connecting the regions with the Earth) have been determined by measuring the Doppler shifts of the 21-cm radiation. A drawing of the Milky Way Galaxy, constructed from 21-cm measurements, is shown in Fig. 13.6. Note the similarity of the spiral structure of our Galaxy to that of the Whirlpool galaxy (NGC 5194), shown in Fig. 1.11.

MOLECULES IN SPACE

In 1963 a new branch of radio astronomy was opened with the discovery that there are *molecules* in interstellar space and that these molecules can be studied through their radio-frequency emissions and absorptions. The first *chemically-bound* substance to be observed in space was the hydroxyl system, OH. Although hydrogen is found everywhere in our Galaxy, OH is found primarily in extremely small regions of space near very hot stars. In these regions the OH abundance is $\sim 10^{-4}$ of the hydrogen abundance, a remarkably high value for a molecular unit in interstellar space.

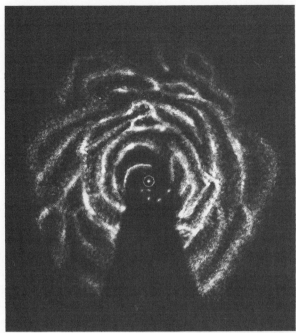

Fig. 13.6 *The spiral structure of the Milky Way as determined from 21-cm measurements of hydrogen concentrations and Doppler shifts. The cross marks the center of the Galaxy and the circled dot in the upper portion of the diagram represents the position of the Sun. The blank region at the bottom is that portion of the Galaxy which is obscured by the dense galactic center.*

Professor Westerhout, University of Maryland

Since the discovery of OH in space, searches for other molecular species have revealed the presence of water (H_2O), carbon monoxide (CO), cyanogen (CN), ammonia (NH_3), hydrogen cyanide (HCN), methyl alcohol (CH_3OH), formic acid (HCOOH), and formaldehyde (CH_2O), all identified on the basis of molecular radio-frequency emissions and absorptions. What is the origin of these space molecules? Do they reveal the presence of life elsewhere in the Galaxy? Or are they only the natural result of the interaction of radiation with the common elements, hydrogen, carbon, oxygen, and nitrogen? Much additional information must be gathered before we will have any tenable answers to these questions.

13.3 *Stellar Evolution*

THE MAIN SEQUENCE

Stars condense from interstellar gases, burn their nuclear fuels, and die an explosive death as supernovae or simply "fade away" and become small, cold collections of nuclear ash. It is fortunate that we can trace stars through their evolutionary history in terms of only two quantities that are relatively easy to measure—the *intrinsic luminosity* and the *color* (which indicates the *surface temperature* of a star). Every star, at every epoch in its life, can be represented by a single point in a color-luminosity diagram. During the life of a star, this representative point traces out a path, and the life and death process of a star can be shown in a simple and graphic way in terms of a curve in a color-luminosity diagram.

The dynamical behavior of a star depends on only two factors—the mass of the material from which the star has condensed and the composition of that material. In the initial phases of a star's life, only the *mass* is important in determining the stellar dynamics. The *composition* influences the sequence of nuclear reactions that take place during later phases.

If we consider stars that are similar in composition to the Sun, then we find that these stars spend most of their lifetimes occupying positions along a color-luminosity curve called the *main sequence* (see Fig. 13.7). The position

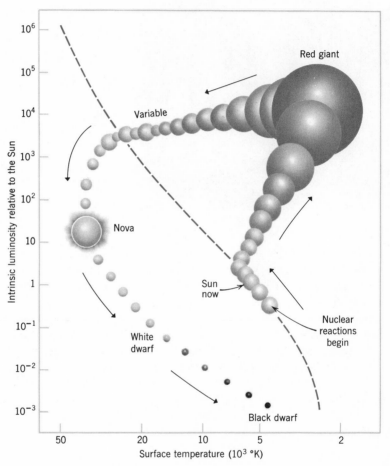

Fig. 13.7 *A simplified diagram of the evolutionary track of a typical star (e.g., the Sun).*

of a star on this curve when nuclear reactions begin to take place in the core and appreciable amounts of radiation are emitted depends on the mass of the star—the more massive stars are hot and bright, whereas the less massive stars are cool and dim.

As a star condenses, the release of gravitational potential energy increases the central temperature until thermonuclear reactions are initiated. This new source of energy causes the contraction process to become slower because of the radiation pressure exerted on the outer layers of the star. Eventually, the rate of generation of thermonuclear energy increases to the point that the outward radiation pressure on any small volume of star material is just equal to the inward gravitational pressure. The star then becomes stabilized in size and in luminosity. A star spends most of its life in this condition and, during this time, the star's representative point moves only a short distance along the main sequence.

EVOLUTION OFF THE MAIN SEQUENCE

A typical star spends 90 percent or more of its lifetime burning hydrogen in its core by the *p-p* chain or CN cycle reactions and simultaneously moving slowly up the main sequence. The Sun has been doing this for 4.5×10^9 years and will continue to do so for another 5×10^9 years before embarking upon the last, violent stages of its life. More massive stars evolve much more rapidly—a star that has a mass of 20–50 solar masses will run through its life cycle in the comparatively brief period of a few million years.

When hydrogen is exhausted in a star's core, the core collapses, the release of gravitational energy causes the central temperature to increase, and helium burning begins. Since a large amount of energy is released in converting helium into carbon, the luminosity of the star increases. But this energy release also means increased radiation pressure on the outer portion of the star's gases. Consequently, the outer layers of the star *expand*. This expansion causes the gases to become cooler and therefore the emitted light becomes *redder*. Thus, the star departs abruptly from the main sequence (see Fig. 13.7). This expansion and reddening continues until the star has a diameter 200–300 times the diameter it had while on the main sequence—the star becomes a *red giant*.[2] As the Sun reaches the red-giant stage, it will first toast the Earth to a cinder (because of the increased energy output) and then engulf the ashes (because of the tremendous expansion). But this catastrophic day of reckoning is ~5 billion years away.

The red-giant phase in the life of a typical star lasts ~10^7 years. Once having reached its maximum size in the red-giant stage, a star evolves rapidly and moves to the left in the color-luminosity diagram. In fact, the time required to evolve from the red-giant stage to the point at which it crosses the main sequence line (see Fig. 13.7) requires only about one percent of the star's lifetime. Thus, the Sun will make this transition in about 100 million years. At this stage in their careers, most stars become dynamically unstable and begin to pulsate, increasing and decreasing in both size and luminosity. Most *variable stars* are found in this region of the color-luminosity diagram.

NOVAE

Up to the point at which the red-giant stage is reached, it is believed that all stars evolve in approximately the same way. (The *rate* of evolution depends on the mass of the star.) During the red-giant and subsequent phases, however, differences in composition become important. As a star evolves from the red-giant maximum and moves to the left in the color-luminosity diagram, the central temperature increases (as does the surface temperature). This means that new thermonuclear reactions can take place,

[2]The red color of the giant star *Betelgeuse* (Orion's right shoulder) is quite apparent to the unaided eye.

and the composition of the star determines which reactions will occur. This is a very complicated stage in a star's life and our theories of post-red-giant evolution are still incomplete. Although we do not understand the details, it seems clear that there are at least two distinct possibilities for the behavior of a star during its terminal phase—massive stars explode and small stars just "fade away."

As a small star (mass < 1.4 times the solar mass) consumes the last of its nuclear fuel, it progresses downward on the color-luminosity diagram (see Fig. 13.7); the luminosity and energy output decrease as gravitational contraction attempts to supply sufficient energy to maintain the temperature. Before substantial cooling has taken place, the star may go through another stage of instability in which there are periodic eruptions that spew stellar material into space. Each such eruption may cause 10^{-4} to 10^{-5} of the star's mass to be ejected into space while at the same time increasing the star's luminosity. A star that exhibits such a surge of light output is called a *nova*.[3] A typical nova will increase in luminosity by a factor of about 10^3 in a period of less than a day. This increased light output persists for a week or two and then declines. Generally, a nova will recur, often many times, until a condition of stability is reached.

As a prelude to entering the region of the color-luminosity diagram occupied by *white dwarfs* (Fig. 13.7), a star may eject still more of its material, but not in as violent a way as a nova. These prewhite-dwarf ejections are probably responsible for the spectacular ring-like gas clouds (called *planetary nebulae*) that are observed to encircle some stars (Fig. 13.8).

When the last of a white dwarf's available energy is radiated away, it cools rapidly, first turning red (a *red dwarf*), and finally becoming a cold, dense cinder of a once-mighty nuclear furnace—a *black dwarf*.

SUPERNOVAE

Stars that are 1.4 or more times as massive as the Sun die a spectacular death. Instead of entering a period of relatively gentle ejections of material, as do the small stars in the nova phase, massive stars expire in a single gigantic explosion. In such a *supernova,* the light intensity can be 10^4 times that of a typical nova. During the period of time in which astronomical events have been recorded, three supernovae in our Galaxy have been visible to the unaided eye. The first was the explosion of 1054 which occurred in the constellation Taurus and was recorded by the Chinese. The second was observed by Tycho Brahe in Cassiopeia in 1572, and the most recent, the supernova of 1604 in Serpens, was described both by Kepler and Galileo. Many other supernovae in other galaxies have been observed telescopically (see Fig. 13.9). In a typical galaxy, supernovae seem to occur at a rate of one every few hundred years.

According to a current theory, supernovae originate in the following way. As helium burning proceeds in the core of a star during the red-giant phase, the central temperature increases. At temperatures of a few times $10^9 °K$, fusion reactions produce nuclei with higher and higher atomic numbers until, finally, the group of elements near iron is formed. The nuclei of these

[3] So called because early observers believed these objects to be *new* stars (they were usually not visible prior to the outburst).

Fig. 13.8 *The planetary nebula NGC 7293 in the constellation Aquarius, photographed with the 200-in. telescope. More than 1000 examples of planetary nebulae are known. The ejection of gaseous material that forms a planetary nebula seems to be a part of the evolutionary process that leads to the formation of a white dwarf star.*

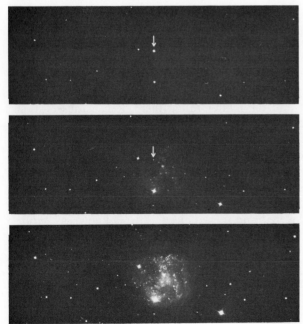

Fig. 13.9 *Three photographs of the 1937 supernova in Virgo. The first photograph (1937), a 20-min exposure, shows the bright star after the maximum intensity had been attained. The second photograph (1938), a 45-min exposure, just reveals the waning star. The last photograph (1942) fails to reveal any trace of the star even though the exposure was 85 min. (The increased exposure time for this photograph has made visible many faint stars and some nebulosity.)*

315

elements have the largest binding energies of all nuclei (see Fig. 12.5), and therefore additional fusion reactions do not release energy. Consequently, the generation of thermonuclear energy in the core can proceed no further. Because the radiation pressure is then insufficient to maintain the stability of the outer layers of the star, a gravitational collapse of the core takes place. The implosion of the core draws in the envelope material of the star, which contains unconsumed nuclear fuels. As this new material is heated, further thermonuclear reactions occur and the material is raised to higher temperatures. This new source of energy reverses the collapse and a catastrophic explosion ensues. During the collapse and subsequent explosion (which takes place during the remarkably short period of only a few minutes), enormous numbers of neutrons are produced and heavy elements are formed by neutron capture reactions. In supernovae explosions, most of the star material, consisting of both light and heavy elements, is ejected into space; this material is then incorporated into stars of the next generation as they condense from interstellar material.

It is now believed that most of the high-energy cosmic rays that continually bombard the Earth are particles that were ejected with extremely high velocities from supernovae explosions. These particles were then trapped in the galactic magnetic field and have eventually found their way to the Earth. The study of the composition of cosmic rays has been of considerable importance in determining the abundance of the elements in the Universe.

NEUTRON STARS AND PULSARS

What remains after a supernova explosion? As long ago as 1934 it was proposed that when a condition of extreme pressure and temperature is attained in the core of a star, electrons can literally be squeezed into nuclei where they combine with protons to form neutrons. Electrostatic repulsion would be removed by such a process and the neutrons would collapse under gravitational attraction into a small superdense ball, so dense, in fact, that the neutrons would be prevented from undergoing the normal decay process. Stars that explode as supernovae appear to satisfy the conditions for neutron formation in their cores, and so it was thought that the remnant of a supernova explosion would be a *neutron star*. But no one knew how to identify a neutron star and the matter rested there. Recent observations, however, have revealed several unusual star-like objects that are probably the conjectured neutron stars.

In 1968 a startling new discovery was announced by a group of radio astronomers at Cambridge University. They had detected a radio source with the unique feature that it pulsates at a rapid rate—approximately once per second. Soon after the Cambridge announcement, several other pulsating radio sources were discovered and these objects became known as *pulsars*.

Thomas Gold of Cornell University proposed a model of pulsars that appears capable of accounting for many of the remarkable properties of these objects. According to Gold, a pulsar is a rapidly rotating neutron star. The pulsed radiation arises in the following way. Electrons and protons are trapped in the enormous magnetic field of the neutron star ($B \sim 10^8$ T!). As the star spins, the magnetic field and the trapped particles spin with the star. At the outer limit of this magnetically confined plasma the particles travel with velocities approaching the velocity of light. Because the motion is circular, the particles are accelerated and therefore emit radiation. The radiation is particularly intense because of the extremely high accelerations of the particles. Another consequence of the relativistic velocities of the particles is the fact that the radiation is emitted almost entirely along the direction of motion of the particles. Since the particles rotate with the magnetic field of the star, the radiation takes the form of a "searchlight beacon" scanning the sky (see Fig. 13.10). Once each revolution, a beam flashes toward the Earth.

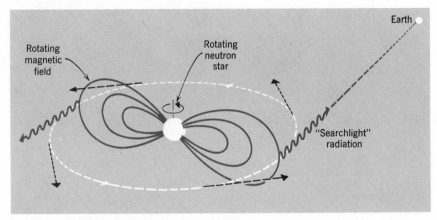

Fig. 13.10 *Gold's model of a pulsar. Charged particles are trapped in the star's magnetic field which rotates with the star. The accelerated particles emit radiation similar to that from a rotating searchlight beacon.*

Pulsating *optical* radiation from a pulsar was first observed in 1969. Measurements made at the University of California's Lick Observatory have shown that the pulsar in the Crab Nebula winks on and off with the same 0.033-sec period found for the radio signal (see Fig. 13.11).

Although the idea of a rotating neutron star seems capable of explaining much of the present accumulation of pulsar data, it is, of course, not certain whether this model will stand the test of further observations. But it is an interesting answer to an unusual astronomical puzzle, one that emphasizes the fact that imaginative new ideas are necessary to account for the vast range of phenomena observable in the Universe.

BLACK HOLES

One of the most spectacular predictions of Einstein's general theory of relativity is that a sufficient mass of cool matter (that is, matter with low kinetic energy) will undergo a continuing gravitational collapse. For example, consider a massive star that undergoes a supernova explosion. If the residual core matter is *hot* and the particles are moving rapidly, then the gravitational contraction can proceed no further than the formation of a neutron star. But if the core matter is *cool,* then the slowly moving particles cannot halt the collapse at a density corresponding to that of nuclear matter (thus forming a neutron star). Consequently, the collapse will continue and

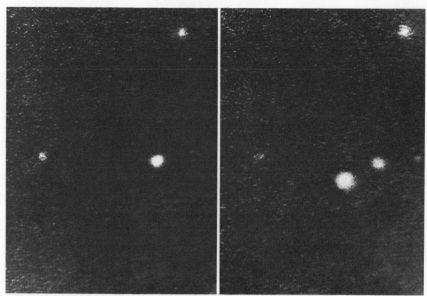

J. S. Miller & E. J. Wampler, Lick Observatory

Fig. 13.11 *Optical radiation from the Crab pulsar—(a)* off *and (b)* on. *The photographs are from a light-amplifying television camera attached to the Lick Observatory telescope. The light signals show the same pulsation property and the same period as the radio signals.*

ever-higher densities will result. The gravitational field in the vicinity of such a collapsing object is enormously intense. In Section 9.5 it was mentioned that light (in fact, *any* electromagnetic radiation) has equivalent mass and therefore reacts to a gravitational field. When a supernova core has collapsed to a state of extremely high density, the gravitational field can be sufficiently intense to prevent any light from leaving the object and it becomes a *black hole,* incapable of emitting any radiation! (See Fig. 13.12.) How could we detect such a peculiar object? There appear to be several ways that offer possibilities. One method would be to observe the motion of another star orbiting the *black hole* (but at a sufficient distance that the star will not be drawn into the *black hole* and disappear); at least one candidate of this type has been identified. Also, a collapsing object will radiate gravitational waves and should be detectable by methods such as that developed by Weber (see Section 9.5). Perhaps *black holes* have already been detected!

It should be pointed out that a *black hole* (sometimes called a *collapsar*) is a truly relativistic object. A neutron star could exist according to the laws of Newtonian dynamics, but a *black hole* is entirely the result of general relativistic effects.

13.4 *Cosmology*

THE BIG BANG THEORY

The idea that the Universe originated in a gigantic explosion (a *big bang*) was first propounded by the Russian-American theorist, George Gamow (1904–1968) in 1948. According to the big bang theory, about 10^{10} years

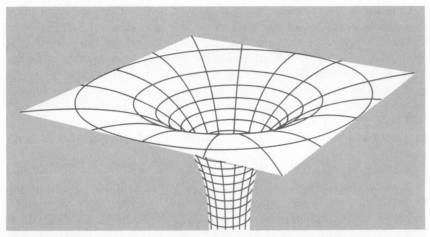

Fig. 13.12 *Schematic diagram of the gravitational potential field around a* black hole.

ago all of the matter and energy now in the Universe were concentrated in a single *fireball* in which the density was $\gtrsim 10^{28}$ kg/m³ and the temperature was $\gtrsim 10^{16}$°K. The radiation pressure was tremendous in this fireball and it expanded outward with explosive rapidity—the *big bang*. Those parts of the fireball that had the greatest relative velocities are now concentrated in the distant galaxies that we see (as they were $\sim 2 \times 10^9$ years ago) receding from us with high velocities. Thus, the general expansion of the Universe results in a natural way from the big bang theory.

Gamow originally proposed that all of the elements in the Universe were formed by nuclear reactions during the first few moments after the big bang. This hypothesis leads to the conclusion that the Universe should everywhere exhibit the same relative abundance of the elements. But we now know from spectral measurements that even in the Milky Way there are wide variations in element abundances among the stars. And, indeed, our current theories of stellar evolution make clear the point that element synthesis is a continuing process and that erupting and exploding stars return the products of thermonuclear reactions to the interstellar medium where they can be incorporated into newly forming stars.

Refinements of Gamow's big bang theory have shown that nuclear reactions did indeed take place in the expanding fireball but that the reactions proceeded no further than the formation of helium. In fact, detailed calculations show that the reactions in the fireball built up helium to an abundance about 10 percent of that of hydrogen. Thus, the big bang theory can account for the observed universal abundance of helium (see Section 13.1).

The most spectacular confirmation of a prediction of the big bang theory has been the recent observation of low-frequency radiation emanating from space and presumably associated with the fireball. In the superdense fireball there was both matter and radiation. The radiation was in the form of enormously energetic γ rays, but as the fireball expanded and cooled, the γ radiation also "cooled" and the photon energies decreased (that is, the wavelengths increased). This radiation still persists in the Universe, but it is now in the form of radio waves, microwaves, and some infrared radiation.

13.4
COSMOLOGY

In thinking about the fireball, we must remember that we cannot place ourselves at some distant position and "view" the fireball expanding toward us. The fireball *is* the Universe, and the Earth (or at least the raw material from which the Earth will eventually be formed) is immersed within the fireball. As the fireball expands, all the rest of the matter in the Universe is moving away from the Earth-to-be (or from any other piece of matter in the fireball). Therefore, the fireball radiation bombards the Earth (*then* and *now*) from all directions. Indeed, any observer in the Universe would detect this radiation arriving at his location in equal amounts from all directions.

Because the fireball has been expanding for $\sim 10^{10}$ years, the original fantastically high temperature has decreased to the point that now, according to the theory, the mean temperature of the Universe should be approximately $3\,^{\circ}\mathrm{K}$. The radiation from a source at $3\,^{\circ}\mathrm{K}$ should have a distribution of wavelengths with a maximum near 10^{-3} m $= 1$ mm. If the big bang theory is correct, there should be two observable effects: (1) The radiation spectrum should have the shape characteristic of a $3\,^{\circ}$-K source and (2) this radiation should be arriving in equal amounts from all directions in space (that is, the radiation should be *isotropic*).

In 1965 the first measurements were made which detected low-energy cosmic radio waves that could be interpreted as radiation from the still-expanding-but-now-cool fireball. Subsequently, sufficient additional measurements were made to establish that the shape of the spectrum is indeed that predicted by radiation theory and corresponds to a temperature of $2.7\,^{\circ}\mathrm{K}$ (see Fig. 13.13). Furthermore, measurements made of the radiation arriving

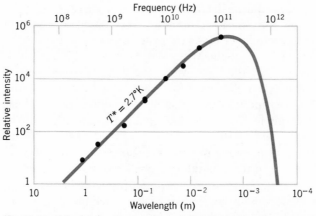

Fig. 13.13 *Measured values of the intensity of low-energy cosmic radio waves (solid circles) compared with the spectrum shape expected on the basis of radiation theory. These measurements indicate that the mean temperature of the Universe is 2.7° K. The fact that the spectrum agrees closely with the theoretical spectrum is considered to be strong evidence in favor of the big bang theory of the origin of the Universe.*

at the Earth from different directions in space have shown that the radiation is isotropic to within a few tenths of one percent.

The big bang theory of the origin of the Universe has therefore successfully passed the crucial observational tests that have so far been devised. But the theory is at present in its formative stages and sophisticated measurement techniques designed to test the theory are still being developed. We can therefore expect new confrontations between theory and experiment to take place at an increasing rate during the next few years. Whether the theory will survive these tests remains to be seen.

THE FUTURE OF THE UNIVERSE

The Universe is now expanding. Will this expansion continue into the indefinite future so that the matter in the Universe will approach a state of being infinitely dilute? Relativity theory makes a definite prediction on this point. According to the theory, there exists a *critical mass* for the Universe. If the actual mass is *less* than this critical value, the mutual gravitational attraction of all the matter in the Universe is insufficient to halt the expansion and the Universe will continue on toward infinite dilution. On the other hand, if the actual mass of the Universe *exceeds* the critical value, gravitational attraction will eventually slow the expansion, stop it entirely, and then reverse the motion. In this situation the Universe is destined for eventual collapse and the fireball will be reformed. The stage will then be set for another big bang and another expansion. That is, the Universe *oscillates* between a condensed fireball phase and a phase of maximum expansion.

Does the Universe contain sufficient mass (in the form of matter and energy) to bring about the oscillation situation? The approximate amount of matter in stars and galactic dust and gases can be estimated by several means. The amount of energy in star light, in magnetic fields in space, in the motion of gas clouds, in cosmic rays, and in neutrinos can similarly be estimated. All of this mass-energy taken together does not equal the critical mass. There is, however, a great uncertainty in this calculation for we do not know how much matter is in *intergalactic* space. The atomic hydrogen in galactic gas clouds can be detected by means of the 21-cm radiation. But intergalactic hydrogen is probably mostly *ionized* (by virtue of the absorption of radiation emanating from galaxies, radio galaxies, and quasars) and therefore will not emit the characteristic 21-cm radiation. X-ray methods are required to detect ionized hydrogen and these experiments are difficult to perform because they must be carried out above the Earth's atmosphere to eliminate absorption effects. Nevertheless, some recent measurements with rockets and satellites indicate the possible existence of appreciable amounts of ionized hydrogen in intergalactic space. If the preliminary interpretation of these measurements is indeed correct, then the amount of intergalactic matter appears to be sufficient to make the total mass of the Universe well in excess of the critical value. If this is so, then we live in an oscillating Universe that goes through a cycle from big bang to expansion to collapse every 8×10^{10} years or so.

Summary of Important Ideas

Stars condense from clouds of gas (and dust) that consist primarily of *hydrogen* and *helium*. Some clouds contain heavy-element material ejected into space by novae and supernovae.

As a star condenses, gravitational potential energy is converted into kinetic energy of the atoms and the temperature of the star increases. When $T \sim 10^7 °K$ and $\rho \sim 10^5 \text{ kg/m}^3$, the *proton-proton chain* reactions begin to take place. These reactions convert hydrogen into helium. The *p-p* chain is the main source of energy in the Sun.

In stars that contain carbon, the *CN cycle* is the main source of energy if $T \gtrsim 2 \times 10^7 °K$.

Heavy elements are formed in stars by *helium burning* and by *neutron capture reactions*. In *supernovae* the neutron density is sufficient to produce the heavy, radioactive elements.

Certain *peculiar galaxies* and *quasi-stellar objects* (quasars) emit enormous amounts of radio energy. The process or processes by which this energy is generated is not known. Nor do we understand the roles that these objects play in the cosmological scheme of things.

All stars spend most of their lives on the *main sequence* generating energy by converting hydrogen into helium by the *p-p* chain reactions or the CN cycle reactions.

Stellar *evolution,* traced in a *color-luminosity diagram,* proceeds by a slow movement along the main sequence, followed rapidly by the red-giant, variable, and eruptive stages. Small stars ($m < 1.4\, m_{\text{Sun}}$) eventually become dwarfs, whereas massive stars ($m > 1.4\, m_{\text{Sun}}$) undergo violent explosions (that is, they become *supernovae*). The more massive a star, the more rapidly the evolution takes place.

Pulsars are believed to be rapidly rotating *neutron stars* that are remnants of supernovae explosions.

The weight of observational evidence at present favors a model of the Universe which "begins" with a gigantic explosion (the *big bang* theory). Galaxies and stars condense from the gases in the expanding fireball. The Universe is now expanding from this explosion. Whether the expansion will eventually cease and the matter in the Universe will collapse on itself and form another fireball or whether the expansion will continue to a state of infinite dilution is still an open question.

Questions

13.1

A large mass of gas, originally more or less spherical in shape, condenses into a galaxy of stars. Explain how this process can result in a galaxy that is disc-shaped, as is our Galaxy. (Consider angular momentum effects.)

13.2

What are the various pieces of evidence that there is matter in the interstellar regions of space?

13.3

Some stars that are still on the main sequence have binary companions that are *white dwarfs*. If the stars were formed at the same time, explain how they can now be so different.

13.4

Spectral lines of the element *technitium* have been observed in the light from certain red-giant stars. The longest-lived isotope of technitium has a half-life of 2×10^6 years. How does this fact support the argument that element formation must take place through nuclear reactions in stars?

13.5

A certain galaxy contains a large amount of hydrogen and helium (but no heavy elements) in gas clouds. Do you expect the stars in this galaxy to be relatively old or relatively young? Explain.

13.6

Consider two galaxies, one of which is very old and the other of which has been formed only recently (on a cosmological time scale!). How would the stars in these two galaxies differ?

13.7

Assume that the oscillating model of the Universe is correct. Will star formation take place during all phases of a cycle of the Universe or will there be periods during which star formation ceases? Explain.

13.8

If the Universe is indeed *oscillating*, will we always observe red shifts from the distant galaxies? Explain.

Problems

13.1

The Sun is moving with a speed of approximately 300 km/sec around the galactic center which is at a distance of 3×10^4 L.Y. About how many orbits of our Galaxy will the Sun make during the remainder of its expected life?

13.2

We know that a typical star radiates enormous amounts of energy and, therefore, that the mass of the star decreases with time. Why, then, do we refer to *the* mass of a star? What fraction of the mass of the Sun has been radiated away during the last 10^9 years? (The Sun's radiation rate has been essentially constant during this time.)

APPENDIX
RADIAN MEASURE

CIRCULAR MEASURE

The most common unit of angular measure is the *degree,* defined to be 1/360 of a complete circle. For some purposes it is more convenient to use another unit, the *radian* (Fig. A.1). We choose a unit of angular measure so that if $\theta = 1$ unit, the length of the portion s of the circumference of

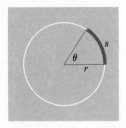

Fig. A.1 *The central angle θ defines an arc of length* s *on the circumference of a circle of radius* r.

the circle (called the *arc*) that is intercepted by the pair of radii defining the angle is just equal to the radius r of the circle. This unit is called the *radian*. Then, we have, in general,

$$s = r\theta \tag{A.1}$$

where θ is measured in radians. The circumference c of a circle is 2π times the radius:

$$c = 2\pi r \tag{A.2}$$

Therefore, the 360° in a complete circle is equivalent to 2π radians. Thus,

$$1 \text{ radian} = \frac{360°}{2\pi} \cong 57°3 \tag{A.3}$$

If we use a more accurate value of π (3.14159265 . . .), we find that 1 radian is 57°2957795 . . . . Also, 1° = 0.0174533 . . . radian.

The fact that $s = r\theta$ provides us with a method of closely estimating distances in certain circumstances. Suppose that a pole is placed vertically

325

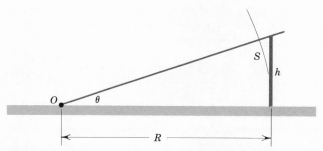

Fig. A.2 *Estimating the height of an object by measuring the angle subtended at O;* h ≅ Rθ.

in the ground a distance R away from an observer located at a point O (Fig. A.2). The observer measures the angle subtended by the pole and finds it to be θ. Mentally, the observer constructs a circle of radius R centered at O and passing through the bottom of the pole. The arc length S is then given by

$$S = R\theta \tag{A.4}$$

Since θ is a small angle, S is approximately equal to the height of the pole h, and we can write

$$h \cong R\theta \tag{A.5}$$

This method of obtaining an approximate value for h will be useful if θ is sufficiently small. Even for θ as large as 20°, the error is only about 4 percent. If θ is a few degrees, the error is usually negligible for most purposes; for $\theta = 1°$, the error is 0.01 percent or 1 part in 10^4.

If we knew h, we could find the distance R to the pole by using

$$\boxed{R \cong h/\theta} \tag{A.6}$$

CHAPTER 1

1.1 $x = n^2, n = 1, 2, 3, \ldots$

1.3   (a) 0.0006   (c) 0.000 0039
       (b) 0.000 000 86   (d) 0.000 000 000 003

1.5   10 sec

1.7   1.04 mi/min

1.9 2.5×10^9

1.11   1 in^2 = 6.453 cm^2; 1 ft^3 = 0.284 m^3

1.13 2.1×10^{22} m

1.15 2.7×10^9 L.Y.

CHAPTER 2

2.3   (a) 30 ft/sec   (b) 10 ft/sec   (c) 18 ft/sec

2.5   The velocities will be the same.

2.7 $t = 7$ sec

2.9   1280 ft/sec; 6400 ft

2.11   27g

2.13   (a) 2 units   (b) 1 unit   (c) 0

2.17   0.6 m/sec

2.19 7.3×10^{-5} rad/sec; 0.47 km/sec

2.21   21 ft/sec; 2.8 sec

CHAPTER 3

3.1   2 m/sec^2

3.3   0.1 N, 0.2 N; 2 m/sec^2

3.5   300 m/sec

3.7   2000 N in the direction opposite to the initial motion.

3.9   ∼6 cm/sec; very likely

3.11 $v = 1.67$ m/sec; absorbed by the Earth

3.13   The block continues its circular motion with the same angular velocity; the chunk moves in a straight line with a speed equal to the original speed of the block (4 m/sec).

CHAPTER 4

4.1 6.7×10^{-9} N

4.3 3.5×10^{32} N

4.5 2.74×10^2 m/sec^2

4.7 -2×10^{-4} C

4.9 2.3×10^{-8} N; $a_e = 2.5 \times 10^{22}$ m/sec^2, $a_p = 1.38 \times 10^{19}$ m/sec^2

CHAPTER 5

5.1 12 m/sec

5.3 1.64×10^5 N

5.5 7.84×10^8 J; 1.18×10^9 J; no, additional PE is made available.

5.7 22.6 m/sec, 10.8 m/sec

5.9 2.25×10^5 J

5.11 1.26×10^9 J; 5×10^3 m/sec

5.13 3×10^6 eV = 3 MeV; the electrons are accelerated toward A.

5.15 $0.47°C$

5.17 6%

5.19 50 kg

CHAPTER 6

6.1 g = 9.8 m/sec^2

6.3 $|\mathbf{E}| = 140$ V/m, $\mathbf{E}$ is directed *toward* the origin; $\Phi_E = -800$ V

6.5 1.6×10^{-14} N

6.7 $E = 1.13 \times 10^5$ V/m, $\Phi_E = 1.13 \times 10^5$ V

6.9 1.92×10^{-16} N

CHAPTER 7

7.1 1.36 A; 81 Ω

7.5 $F_{max} = 9.5 \times 10^{-12}$ N, when electron moves perpendicular to $\mathbf{B}$; $F_{min} = 0$, when electron moves parallel to $\mathbf{B}$.

7.7 5.35×10^{-4} T

7.9 1/6 MeV

7.11 $(BR)_p = (BR)_\alpha$

7.13 $R = 10^{19}$ m $\cong$ 1% of size of our Galaxy

7.15 0.9 T

CHAPTER 8

8.1 10^{-2} N

8.3 The signal propagating through water will be heard first; 7.1 sec

8.5 50

8.7 0.1 mm

8.9 $\Delta x = 0.3$ m

8.11 1.5×10^{21} Hz; gamma ray

CHAPTER 9

ANSWERS TO ODD-NUM-BERED PROB-LEMS

9.1 Only the second explosion could have been caused by pushing the firing button.

9.3 $v = (\sqrt{3}/2)c = 0.866c = 2.59 \times 10^8$ m/sec

9.5 1.8c (observer on Earth); 0.994c (observer on spaceship)

9.7 0.5 m

CHAPTER 10

10.1 4.1 eV

10.3 2330 Å

10.5 12.4 eV = 1.98×10^{-18} J

10.7 3×10^{18} Hz; 12.4 eV

10.9 Electron: 1.39×10^{-6} sec; proton: 2.55×10^{-3} sec; photon: 3.3×10^{-8} sec

10.11 1.8×10^{-10} m

10.13 6.6×10^{-34} kg-m/sec

10.15 1.33×10^{-24} kg-m/sec; 6 eV; because the *total* energy remains constant

10.17 2×10^{-10} sec

CHAPTER 11

11.1 4.2×10^{-14} m

11.3 910 Å

11.5 Only 12.1, 10.2, and 1.88 eV

11.7 The K shell will fill at $Z = 4$ (Be), and the L shell will fill at $Z = 20$ (Ca).

11.9 Above 2 eV. (Light with frequency and energy higher than for yellow light is absorbed by raising electrons into the conduction band.)

CHAPTER 12

12.1 β^+ decay

12.3 63.6 AMU (actually, 63.54 AMU)

12.5 C^{12}; 28.18 MeV

12.7 C^{12}

12.9 16,700 years ago (~15,000 B.C.)

12.11 18.6 keV

12.13 1877.6 MeV; pions and photons

CHAPTER 13

13.1 20

ANSWERS TO
ODD-NUM-
BERED PROB-
LEMS

INDEX